Nanoscience

Volume 4

A Specialist Periodical Report

Nanoscience

Volume 4

A Review of Recent Literature

Editors
P. John Thomas, *Bangor University, UK*
Neerish Revaprasadu, *University of Zululand, South Africa*

Authors
Mohammad Al Kobaisi, *RMIT University, Australia*
Anuradha, *RMIT University, Australia*
Ramjee Balasubramanian, *Old Dominion University, USA*
Sean N. Baxter, *Bangor University, UK*
Sheshanath V. Bhosale, *RMIT University, Australia*
Jack R. Brent, *The University of Manchester, UK*
Gilberto Casillas, *University of Wollongong, Australia*
Francis Leonard Deepak, *International Iberian Nanotechnology Laboratory, Portugal*
R. Dunpall, *University of Zululand, South Africa*
Akhil Gupta, *Deakin University, Australia*
Duong Duc La, *RMIT University, Australia*
David J. Lewis, *The University of Manchester, UK*
Karthik Ramasamy, *UbiQD, LLC, USA*
Anushri Rananaware, *RMIT University, Australia*
N. Revaprasadu, *University of Zululand, South Africa*
Richard A. Taylor, *The University of the West Indies, Trinidad and Tobago*
Aleksander A. Tedstone, *The University of Manchester, UK*
P. John Thomas, *Bangor University, UK*

ROYAL SOCIETY OF CHEMISTRY

THE QUEEN'S AWARDS FOR ENTERPRISE: INTERNATIONAL TRADE 2013

ISBN: 978-1-78262-159-1
PDF eISBN: 978-1-78262-035-8
EPUB eISBN: 978-1-78801-249-2
DOI: 10.1039/9781782620358
ISSN: 2049-3541

A catalogue record for this book is available from the British Library

The Royal Society of Chemistry is a charity, registered in England and Wales, Number 207890, and a company incorporated in England by Royal Charter (Registered No. RC000524), registered office: Burlington House, Piccadilly, London W1J 0BA, UK, Telephone: +44 (0) 207 4378 6556.

Visit our website at www.rsc.org/books

Printed in the United Kingdom by CPI Group (UK) Ltd, Croydon, CR0 4YY, UK

Preface

DOI: 10.1039/9781782620358-FP005

Welcome to the fourth Royal Society of Chemistry SPR on Nanoscience. Research in the area of Nanoscience continues to burgeon, attracting interest from all areas of physical and medical sciences. The area is truly at the forefront of worldwide research efforts. Concomitant with this status, is an increase in the volume of outputs and scale of activity. In this context, gaining a perspective of developments is challenging for both new comers to the field and seasoned veterans. We have attempted to address this challenge and provide an overview of the developments in the area, focusing specifically on the past few years. Chapters in this volume cover developments in 2D materials, transmission electron microscopy, cancer treatment, ligand design and synthesis, quantum dot solar cells, luminopphores as well as synthetic strategies for nanocrystals. We very much hope you enjoy reading the book.

P. John Thomas (Bangor)
Neerish Revaprasadu (Zululand)

Author biographies

DOI: 10.1039/9781782620358-FP006

Dr Mohammad Al Kobaisi, born in 1971 in Iraq, awarded his MSc in Physical Chemistry from Shahid Beheshti University in 1999, and PhD under the supervision of A/Prof. Colin Rix and Professor David Mainwaring from RMIT University in 2007 investigating the design of new selective adsorbent polymers for sensing application. Since 2007 he continued his research career working in Professor David Mainwaring team in designing drug delivery and vaccine delivery systems in collaboration with the pharmaceutical industry. His general research interest remains in molecular design and in developing nano- and micro-functional materials as vaccine delivery systems.

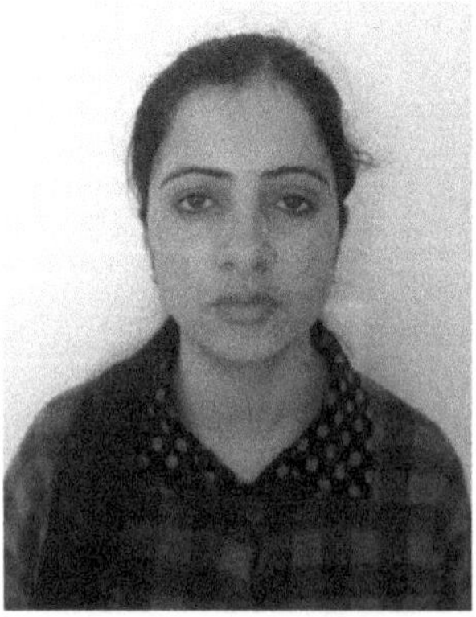

Anuradha obtained her Master's degree in chemistry at Jamia Hamdard University, New Delhi, India. Now she is pursuing PhD degree at RMIT University, Melbourne, Australia with Dr Sheshanath Bhosale. Her interest involves synthesis of Aggregation Induced Emission molecules for application in supramolecular chemistry mainly.

Ramjee Balasubramanian graduated with a B.Sc. (Chemistry) from Loyola College, Madras and obtained his M.S. (Chemical Sciences) and PhD (Organic Chemistry) from the Indian Institute of Science, Bangalore, India. After postdoctoral stints in the laboratories of Prof. Alexander Wei (Department of Chemistry, Purdue University, IN) and Prof. Royce W. Murray (Department of Chemistry, University of North Carolina, Chapel Hill, NC), he joined Old Dominion University as an Assistant Professor. Currently, he is an Associate Professor and his research interests are in the general areas of synthesis, functionalization and applications of polymeric nanocapsules and inorganic nanoparticles.

Sean N. Baxter obtained his MChem from the University of Wales, Bangor, in 2012. He secured a PGCE and spent time teaching before returning to Bangor University, where he is currently a Senior Demonstrator, pursuing his PhD. His research, under the supervision of Dr P. John Thomas, is into the synthesis and uses of thin film nanocrystals. He is a member of the Royal Society of Chemistry.

Sheshanath V. Bhosale (Shesh), born in 1976 in Indral Tq. Devoni, completed his M.Sc. in chemistry from the Udgiri college Udgir (S.R.T.M.U. Nanded) in 1999. He then worked as a project assistant at NCL, Pune, before moving to the Freie University Berlin, Germany, where he received his PhD (*Magna Cum Lauda*) in supramolecular chemistry under the supervision of Prof. J. H. Fuhrhop in 2004. Dr Bhosale pursued his post-doctoral studies with Prof. S. Matile at University of Geneva, Switzerland, under the auspices of a Roche Foundation Fellowship. This was followed by a stay at Monash University, Australia, for five years as an ARC-APD Fellow. Shesh received a prestigious Future Fellowship (2011) from the Australian Research Council and is now working at the School of Science, RMIT University, Melbourne, Australia. Shesh is interested in design and synthesis of π-functional materials, especially small molecules, for sensing, biomaterials and supramolecular chemistry applications.

Jack R. Brent received his M.Chem. from the University of Manchester in 2012 and has recently completed his PhD under the supervision of Paul O'Brien CBE FRS. His doctoral studies focussed on the production and examination of 2D materials, in particular, ultrasonically exfoliated phosphorene and 2D tin(II) sulfide and the solventless growth of MoS2 quantum dots. He is currently a Postdoctoral Research Associate at the University of Manchester, working on the production of 2D field effect transistors with Professor Brian Derby.

Gilberto Casillas is currently a Transmission Electron Microscopist at the Electron Microscopy Centre at the University of Wollongong, Australia. He received his PhD in Physics from the University of Texas at San Antonio, USA, in 2013. His main area of research is focused in the implementation of Aberration-Corrected S/TEM in combination with Electron Energy Loss Spectroscopy in the characterization of nanomaterials for a wide range of applications, as well as in the study of fundamental physical phenomena at the atomic scale. He also focuses on *in situ* TEM experimentation to explore the dynamics of deformation mechanisms at the nanoscale.

Francis Leonard Deepak is a group leader in the Department of Advanced Electron Microscopy, Imaging and Spectroscopy at the International Iberian Nanotechnology Laboratory (Braga, Portugal). He received his PhD in 2005 from Jawaharlal Nehru Centre for Advanced Scientific Research, Bangalore, India. Subsequently, he was a postdoctoral fellow (2006–2008) at the Weizmann Institute of Science, Israel. His broad area of research is focused on the use of advanced electron microscopic and spectroscopic techniques for the study of materials/nanomaterials for various applications, as well as in the study of nanoscale phenomena. He also employs the Focused Ion Beam (FIB) technique extensively towards the investigations of nanodevices.

Dr Rekha Dunpall obtained her PhD in Biochemistry from the University of Zululand, South Africa in 2016. Her research interest lies in nanotoxicology, biosafety and the development of biocompatible nanomaterials for drug delivery and bio-imaging applications. She has presented several scientific papers at accredited conferences such as MAM-14, NanoAfrica and BioNanoMed and has published several articles in ISI journals. She is currently involved in developing and optimising *in vivo* animal models for anticancer and drug delivery research.

Akhil Gupta completed his PhD in 2014 from the Department of Materials Engineering, Monash University, Clayton, Victoria, Australia. He is currently employed as a Research Fellow at Deakin University, Waurn Ponds, Victoria Australia. He is also a visiting Research Fellow at the CSIRO Manufacturing, Clayton, Victoria Australia. He is involved in fundamental and applied research in the areas of organic electronics, solar cells in particular, photonics and drug discovery. He has additional research activities in the areas of supramolecular assembly for light harvesting, device fabrication, and molecular gels for sensing and photovoltaic applications.

Duong Duc La obtained his MSc degree in Nanotechnology of hybrid materials for superhydrophobic surface and hydrogen sensing at HanYang University (Korea) in 2010. Now he is pursuing his PhD degree at RMIT University (Australia) in the group of Dr Sheshanath Bhosale. In his study, he mainly focuses on fabricating hybrid materials of inorganic and organic (supramolecules) compounds *via* novel approach of self-assembly for photocatalysis, sensing and electronics devices. He is also working on the fabrication of graphene-based materials for photovoltaic and adsorption applications.

David J. Lewis (ORCID iD: 0000-0001-5950) is a Lecturer in Functional Ceramic and Inorganic Materials at the University of Manchester, U.K. He received M.Sci. and PhD degrees at the University of Birmingham, U.K. in 2002 and 2006 respectively. Throughout his career DJL has developed interests in nanomaterials of all dimensions for a range of applications. He is a member of both the Royal Society of Chemistry (MRSC) and the American Chemical Society (ACS). In March 2016 DJL was named by the Royal Society of Chemistry as being in the top 10% of highly cited authors in their 'General Chemistry' portfolio of journals.

Dr Karthik Ramasamy is Director of Chemistry at UbiQD LLC, and leads research and development on synthetic methods and scale-up. He received his BSc and MSc in Chemistry from Bharathidasan University, Tiruchirappalli, India. Dr Ramasamy received his PhD in Chemistry under the supervision of Professor Paul O'Brien FRS (founder of Nanoco) at the University of Manchester, UK for a thesis titled "New Molecular Precursors for Metal Sulfides" in 2010. Before joining UbiQD in 2016, Dr Ramasamy held post-doctoral positions at the University of Alabama and Los Alamos National Laboratory. Over the past eight years, his research has mainly focused on developing new synthetic methods for metal chalcogenide nanocrystals. He has more than 50 publications and over 1000 citations relating to quantum dots and is an expert at optimizing and scaling up the synthesis alloyed metal sulfides quantum dots.

Anushri Rananaware received her M.Sc. degree in Organic Chemistry from Fergusson College, Pune, India in 2007. Then she started working in R & D department of Sai Life Sciences Pvt. Ltd, India for 6 years until 2014. Now she is carrying her PhD degree under the guidance of Dr Sheshanath Bhosale at RMIT University, Melbourne, Australia. Her research is mainly focusing on the synthesis of novel AIE materials for various supra-molecular applications.

Neerish Revaprasadu is a professor of Chemistry and SARChI Chair holder in Nanotechnology. He obtained his BSC(Hons.) from University of Natal in 1993 and PhD from Imperial college, London in 2000. He started as a Senior Lecturer at Unizulu 2000, Associate Professor in 2004 and full professor in 2008. His research interest is in the synthesis of semiconductor nanomaterials. He has worked extensively on cadmium, lead, copper, iron and bismuth chalcogenide materials. He also has worked on applications of gold core shell nanoparticles for drug delivery applications. He has co-authored 150 articles in peer reviewed journals and supervised more than 25 post-graduate (M.Sc and PhD) students. In 2014 he was elected as member to the South African Academy of Science (ASSAF).

Richard A. Taylor holds a PhD in Chemistry and a BSc in Pure and Applied Chemistry from The University of the West Indies (UWI) Mona Campus, Jamaica. He has been twice visiting scholar at the Schools of Chemistry and of Materials, University of Manchester, UK, working with Professor Paul O'Brien FRS. He is currently Lecturer in Materials Chemistry at The UWI, St. Augustine Campus, Trinidad. There, through a small research group of PhD students he has focused his research on optoelectronic materials including, novel metal-organic liquid crystals, chalcogenide semiconductor thin films and nanomaterials (quantum dots) and luminescent metal-organic frameworks.

Aleksander A. Tedstone received his M.Chem. from the University of Manchester in 2013, and continues to study at the same institution as a doctoral researcher working in the field of inorganic chemistry and materials synthesis under the supervision of Prof. Paul O'Brien CBE FRS and Dr David J. Lewis. He is interested in the modification and application of two-dimensional materials.

P. John Thomas obtained his PhD from JNCASR, India working with Profs. C. N. R. Rao, FRS and G. U. Kulkarni. He carried out post-doctoral research in Bangalore and in Manchester with Prof. Paul O'Brien, CBE, FRS, FREng. He was then a RCUK Research Fellow in Manchester, before taking up post as a Lecturer of Physical Chemistry and Bangor. He is a Fellow of the Higher Education Academy and a member of the Royal Society of Chemistry.

CONTENTS

Cover

Preface v

Author biographies vi

Role of ligands in the synthesis of bi- and multi-metallic nanocrystals 1

Ramjee Balasubramanian

1 Introduction 1
2 Ligand effects 4
3 Influence of coordination ligands on metal precursors 8
4 Influence of linear surfactants 12
5 Influence of macrocyclic surfactants 21
6 Conclusion 25
References 26

Recent highlights in advanced transmission electron microscopy techniques: applications to nanomaterials 29

Francis Leonard Deepak and Gilberto Casillas

1 Introduction to aberration corrected S/TEM 29
2 AC-TEM imaging of carbon nanomaterials and related hybrid structures 31
3 Nanoparticles 34
4 1D Nanomaterials 37

5 2D Materials 39
6 Beam sensitive materials 42
7 3D Techniques 43
8 *In situ* electron microscopy 45
9 Conclusions 49
Acknowledgements 50
References 50

Gold fabricated core–shell nanoparticles as innovative cancer therapeutic strategies to improve drug delivery 57
R. Dunpall and N. Revaprasadu

1 Introduction to cancer therapeutics 57
2 Gold based core–shell nanoparticles for cancer drug delivery 59
3 Nano-drug formulations for cancer drug delivery 66
4 Pharmaceutical formulation and drug development 69
5 Future perspectives for gold based core–shell nanoparticles 71
Acknowledgements 71
References 71

Supramolecular chemistry of AIE-active tetraphenylethylene luminophores 75
Duong Duc La, Anuradha, Akhil Gupta, Mohammad Al Kobaisi, Anushri Rananaware and Sheshanath V. Bhosale

1 Introduction 75
2 AIE-active TPE luminogen 76
3 General syntheses and reactivity 78
4 Self-assembly of PDI, NDI and porphyrin bearing TPE moieties 78
5 TPE moieties bearing macromolecules 85
6 Chiral assembly 89
7 Applications 95
Acknowledgements 102
References 102

A review of two-dimensional nanomaterials beyond graphene 108
Aleksander A. Tedstone, Jack R. Brent and David J. Lewis

1 Introduction 108
2 Two-dimensional (2D) nanomaterials 109

3 Heterostructures 128
4 Conclusions and outlook 130
Notes and references 130

Colloidal quantum dots solar cells **142**
Richard A. Taylor and Karthik Ramasamy

1 Introduction, historical background of colloidal quantum dots 142
2 Semiconductors: dimensionality and quantum confined structures 143
3 Optoelectronic properties of CQDs 146
4 CQD chemical properties 149
5 Doped CQDs 153
6 CQD synthesis 154
7 Thin film fabrication 156
8 Electrical properties of colloidal quantum dots 158
9 Solar cell device physics and performance 161
10 Conclusion and perspective 164
References 164

The role of ligands in mediating charge transport in nanocrystalline medium **169**
Sean N. Baxter and P. John Thomas

1 Introduction 169
2 Coulomb blockade and single electron tunnelling 169
3 Charge transport in arrays of nanocrystals 172
4 Ligands and charge transport 175
References 182

Role of ligands in the synthesis of bi- and multi-metallic nanocrystals

Ramjee Balasubramanian
DOI: 10.1039/9781782620358-00001

Wet-chemical methods have enabled the size-, shape- and composition-controlled synthesis of bi- and multi-metallic nanoparticles with varying degrees of success. Among several variables involved in colloidal synthesis, coordination ligands surrounding the metal prior to the generation of nanoparticles and ligand surfactants eventually stabilizing the nanoparticles have been known to play a major role in dictating the reduction kinetics and the size- and shape- of the resulting nanoparticles. This review will discuss some of the recent examples of such ligand effects in the synthesis of bi- and multi-metallic nanoparticles.

1 Introduction

Metals such as Au, Cu, Ag, Pb, Sn, Fe and Hg, collectively referred to as "metals of antiquity" were discovered long back and used by a number of ancient civilizations.[1] Alloys are solid solutions of various metals or metals and non-metals. They have been known since the Bronze and Iron ages and continue to play a major role in our day to day life. Bronze, primarily made of Cu and Sn, in reality is a multimetallic system also containing smaller amounts of Ni, Fe, Pb, As, Co, Sb, S *etc.*, whose exact composition was dependent on the geographical region and the time period.[2] The alloying of iron with carbon enabling the production of oxidation resistant steel and cast iron has also been known for a very long time. An investigation of various steel blades and other weaponry from the Middle ages revealed that Turkish (Ottoman empire) and Italian (Middle ages) steel contained the lowest amount of carbon, Japanese steel contained an intermediate amount while Indian steel (from the Mughal period) had the highest amount of carbon.[3] Such alloys with varying compositions show significant variation in their physical properties and even today research efforts are directed towards optimizing the composition of different alloys for various applications.

There is currently a lot of interest in the synthesis and study of nanomaterials, *i.e.*, those with at least one dimension in the 1–100 nm size range, due to their unique electronic, optical, magnetic, catalytic and chemical properties. A variety of nanomaterials comprising non-metals, metals and their alloys have been widely investigated. The study of alloy based metallic systems is particularly fascinating as their physical and chemical properties and ensuing applications can be modulated by varying their size, shape, elemental composition, and surface elemental distribution.[4–7] When compared to the investigations on bimetallic nanoparticles, tri- and multi-metallic nanoparticles are relatively

Old Dominion University, Department of Chemistry and Biochemistry, 4541 Hampton Blvd., Norfolk, VA 23529, USA. E-mail: bramjee@odu.edu

unexplored[4] and are currently attracting a lot of interest. Bi- and multi-metallic nanoparticles hold promise in wide ranging areas[7] from biology to material science in various applications including catalysis,[8] electro-catalysis[9,10] sensing,[11] and multimodal imaging.[12]

In principle, when two metals are mixed they could form a variety of distinct architectures based on their mixing pattern.[4] They could be core–shell segregated alloys, entirely segregated alloys, multishell nanoalloys and randomly or orderly mixed alloys (Fig. 1). Certain inherent parameters of the constituent metals such as bond strength, surface energies, size and electronegativity dictate the formation of various architectures. For example, if the bond between the two metals is stronger than the homo-nuclear bonds, mixing will be favored. Metals with lower surface energy or larger sizes will tend to occupy the surface of these nanoparticles. In addition to magic sizes, alloy nanoparticles may also offer magic compositions.[4] It is worth noting that the mixing pattern strongly depends on the preparation conditions, composition and dimension of the bimetallic nanoparticles.[13]

There is currently a lot of interest in the shape controlled synthesis of bimetallic nanoparticles.[14] In general, alloy nanoparticles can adopt either single crystalline geometric structures such as octahedra or truncated octahedra or non-single crystalline compact structures such as icosahedra, decahedra, polytetrahedra and polyicosahedra.[4] The efficient packing in non-single crystalline structures leading to nonoptimal interatomic distances causes some internal strain, which will not favor the formation of such geometrical structures with larger dimensions.

Similar to the synthesis of monometallic nanoparticles, the preparation methods for multimetallic nanoparticles can be grouped as either "top-down" physical methods or "bottom-up" wet-chemical methods. This review will primarily focus on select wet-chemical methods, which allow precise control of the size and shape of the nanoparticles without involving specialized equipment. Preparation methods of bi-[4,7,15] and multi-[9,11,12] metallic nanoparticles can be generally grouped as (a) co-reduction[9,11] and (b) successive[12] reduction approaches. While core–shell architectures can be prepared by seed-mediated approaches, random alloy

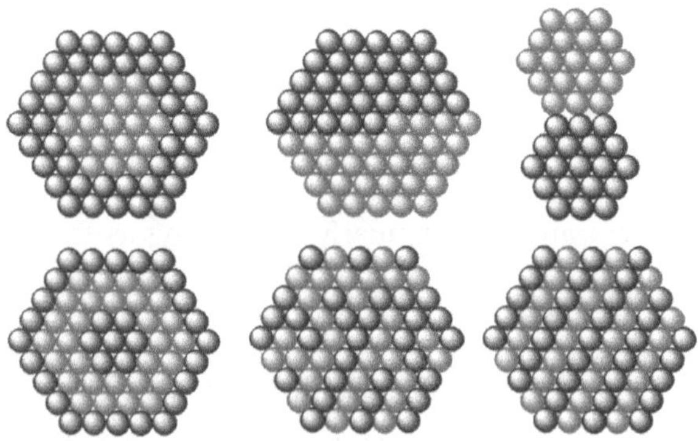

Fig. 1 Distinct bimetallic architectures resulting from various mixing patterns (based on ref. 4).

preparation typically involves co-reduction approaches.[7] In addition, alloy nanoparticles can also be generated by thermal decomposition of organometallic precursors and a variety of methods involving electrochemical, photochemical, sonochemical, biosynthesis, and radiolysis approaches.[4,6,7]

In the co-reduction approach, appropriate mixtures of metal precursors are reduced in the presence of a reducing agent.[4,9,11,15] When compared to seed mediated approaches, this approach offers the simplicity of a one-pot reaction. For example, when palladium and gold salts are co-reduced, given the bond energies of 218.6 ± 6 kJ mol^{-1}, 143 ± 21 kJ mol^{-1}, >136 kJ mol^{-1} for Au–Au, Au–Pd and Pd–Pd bonds respectively, segregation can be predicted.[6] Given the surface energies and sizes of 1.506 J m^{-2} and 144 pm for Au and 2.003 J m^{-2} and 137 pm for Pd, a Pd core–Au shell could be predicted.[6] Indeed molecular dynamics simulations have supported the formation of Pd core–Au shell nanostructures.[4] However, several co-reduction approaches exclusively[16] yield Au core–Pd shell structures which was explained on the basis of reduction kinetics and reduction potential. In general, in the co-reduction approaches the standard reduction potential (SRP) of metal precursors employed can play a pivotal role in determining the segregation or mixing of metals.[17] The metal with the more positive SRP, *i.e.* more noble, easily reducible metal precursor will be reduced first, leading to the formation of the core. The reduction of the second metal precursor (with a lower SRP) as a shell on such cores will lead to the formation of core–shell architectures. From a practical point of view, relative concentration can be effectively employed to circumvent the core–shell formation dictated by SRPs. For example, Sun and coworkers showed that the reaction of $AgNO_3$ (4 mmol) and $HAuCl_4$ (0.2 mmol) at 120 °C in the presence of oleylamine generated alloy nanoparticles instead of the expected core–shell architectures.[18] Further, they showed that the exact alloy composition depended on the duration of the reaction, with $Au_{0.52}Ag_{0.48}$ formed after 1 h of heating and $Au_{0.39}Ag_{0.61}$ at 2 h. On the other hand, metals with comparable SRPs will lead to the formation of alloyed nanostructures.[17] The difference in SRPs between the metal and the reducing agent can also impact the rate of the reduction and the dimensions of the resulting nanoparticles. pH of the reaction medium can also influence the SRP of the metals and thereby influence the formation of nanoparticles.[17]

The growth of anisotropic nanostructures under co-reduction approaches could be classified as (a) continuous growth or (b) crystallite coalescence.[14] In the continuous growth approach, the metal atoms generated upon reduction of metal salts, grow into nuclei, and further into uniform alloy nanoparticles. In this approach, there is a need to control both nucleation and growth of the nanoparticles. The crystallinity of the nuclei generated is controlled by the reduction rate, and their subsequent growth into shaped nanoparticles depends on the select passivation of certain facets by capping agents present in the medium. The growth is restricted on such stabilized facets. Nuclei generated from the metal salts, coalesce into one dimensional or dendritic structures under crystallite coalescence route. When rapid reduction of the metal precursors leads to the formation of a number of nuclei with high surface energy whose subsequent growth is limited by the availability of the metal atoms, they can coalesce in to one large nanoparticle, whose shape is dictated by the growth rate of different facets. Again, the capping agents can control the growth rate of certain facets, and regulate the facet-oriented coalescence process.[14]

In the seeded growth approach, the nucleation and the subsequent growth stages of nanoparticles are temporally separated. The deposition and growth of the second metal on a seed nuclei will depend on the difference in SRPs between the seed metal and the metal to be deposited, lattice match or lack thereof, bond strength between the two metals and finally the facet-specific binding ligands or surfactants.[14] If the SRP of the second metal is more than the SRP of the core metal, then galvanic exchange reaction can occur. For example, Murray *et al.* have shown that thiolate monolayer protected clusters with a Ag, Cu or Pd nanoparticle cores can react with Au–thiolate yielding Au containing bimetallic nanoclusters.[19] In this approach, the added metal ions adsorbed on the seed nanoparticles get reduced by the oxidation of the nanoparticle seeds, and again capping agents can influence the reduction and oxidation sites.[14,19] If the SRP of the metal to be deposited is less than the SRP of the core metal, then seed mediated growth could occur. For example, silver acetate can be reduced on Au nanoparticle surface to yield core–shell architectures.[20] If there is a minimum lattice mismatch between the two metals and if they can bond strongly, it will lead to epitaxial growth.[14] The shape controlled synthesis of nanostructures can proceed simultaneously *via* one or more of the pathways described above. Currently, seed-mediated co-reduction approach is emerging as an important tool in the synthesis of anisotropic nanoparticles.[21] In this approach, two metals are co-reduced in the presence of a pre-formed seed whose symmetry is transferred to the final nanoparticles.

Current synthetic advances have enabled the synthesis of bi- and multi-metallic nanoparticles with distinct morphologies including single-crystalline and twinned polyhedrons, nanodendrites, nanowires, nanorods, multipods, and hollow architectures.[7,9,14,21–35] The key to achieving such shape control depends on a number of experimentally controllable parameters such as choice of metal precursors, stabilizing surfactants, reductants, reaction temperature and media.

2 Ligand effects

Ligands coordinating to the metal precursors can play several important roles during the synthesis of nanoparticles.[36–38] Metal halides are very common precursors employed in both aqueous and two-phase syntheses of nanoparticles. The SRPs of silver salts containing chloride, bromide and iodide are 0.2223, 0.00713, and −0.1522 respectively, which can be used to modulate the reduction rate.[37] The kinetics of nanoparticle growth also depends on the availability of the metal ion, *i.e.*, halides can effectively control the amount of metal ions in the solution. The solubility and the rate of reduction of various gold halides follows the same order: $(AuI_2)^- < (AuBr_2)^- < (AuCl_2)^-$.[37] The stability of the metal precursors which varies with the nature of the coordinating ligand also determines the temperature at which the onset of the nucleation occurs as demonstrated in the case of rhodium nanoparticle synthesis.[38] While $RhCl_3$ is slightly less stable than $RhBr_3$, $Rh_2(TFA)_4$ bridged by a bidentate trifluoroacetate ligand is endowed with exceptional stability requiring higher temperatures.

Such ligands can also determine the morphology of the resulting nanocrystals.[37,38] In the case of Au nanoparticles the binding strength of various halides increases in the following order: $Cl^- < Br^- < I^-$.[37] The binding of ions such as halides to growing metal nanoparticle surfaces can lead to the formation of a physical barrier and thereby retard subsequent metal deposition. Noble metal nanoparticles adopt a face centered cubic lattice and their different crystal planes have different energies.[39] While the surface energies (γ) of fcc metals follows the trend: $\gamma_{(111)} < \gamma_{(100)} < \gamma_{(110)}$, they can be modified by capping ligands or surfactants leading to the formation of nanostructures that would not be thermodynamically favored.[14,38] In the synthesis of rhodium nanoparticles,[38] the selective adsorption of Br^- to (100) planes of rhodium leads to the formation of nanocubes. On the other hand, both TFA^- and Cl^- adsorb to (111) surfaces and lead to the formation of icosahedra and triangular planar plates respectively.[38] In general, Br^- and I^- are known to selectively stabilize the (100) facets of Pt, Pd, Ag, Au and their alloys.[14]

Halides can also undertake additional tasks during the synthesis of nanoparticles in addition to the above described roles, *i.e.* their role in (a) modulating the metal ion's SRP, stability, solubility *etc.* and (b) stabilization of select facets of nuclei/nanoparticles. Halides can also modulate silver under potential deposition, interact with ancillary reagents *etc.* and this has been reviewed elsewhere.[37,40] In addition, they can promote oxidative etching in the presence of oxygen and etch away twinned structures,[14] and initiate, facilitate and direct galvanic exchange reaction.[41] On the other hand, ligands such as citrate ions or citric acid can block the oxidative etching of multiply twinned particles and allow preparation of icosahedra shaped nanoparticles.[42]

To prevent their agglomeration to thermodynamically stable bulk metal, during the synthesis and thereafter nanoparticles are often stabilized with various surface passivating agents, which may be categorized as electrostatic, steric or electrosteric stabilizers.[43] Among steric stabilizers, ligand surfactants, *i.e.*, those that can coordinate to the metal nanoparticle surfaces, bestow metal nanoparticles with enhanced stability against irreversible flocculation or precipitation.[43,44] Thiol based surfactants have been extensively used for the passivation of various metal nanoparticles,[45] though other surface active functional groups[46] such as phosphine, amine, carboxylic acid, isocyanide, pyridone *etc.* have also been employed. Additionally, a variety of polymeric and dendritic surfactants have also been used for stabilizing nanoparticles.[47–49]

Ligand surfactants can exert substantial influence in determining key physical and chemical properties. Stronger binding ligands which can effectively stabilize the nanoparticles often lead to undesirable side effects such as reduced catalytic activity[47,50,51] and dampening of optical properties.[52,53] Hence nanoparticles are often passivated with weakly-binding surfactants like anions, polymers or amine surfactants.[47] On the other hand, the influence of the strongly interacting organic ligands on nanoparticles are not always detrimental. Despite the somewhat reduced conversion (40% of the unmodified catalyst) the modification of Pd with alkanethiol led to improved selectivity during the hydrogenation of 1-epoxy-3-butene.[54]

Ligands and surfactants have been known to show varying affinity for and stabilization of different nanoparticles. While tris(hydroxymethyl) phosphine oxide (THPO) stabilized gold nanoparticles have been known to be unstable, THPO stabilized platinum nanoparticles remain stable.[55] Recently, Rodionov and coworkers exploited the differential binding affinity of the surface passivating ligand towards the constituent metals in influencing the catalysis of FePt bimetallic nanoparticles (Fig. 2).[56] Unlike literature reports which often employ both oleic acid and oleylamine as stabilizers, they synthesized FePt nanoparticles exclusively in the presence of oleic acid. Not surprisingly, they noted that such nanoparticles were more prone to aggregation than those prepared in the presence of both oleic acid and oleylamine. Also, oleic acid could be readily exchanged with other fluorous carboxylic acids, which bind strongly to the surface Fe atoms exclusively while weakly binding and preserving the catalytic activity of surface Pt atoms (Fig. 2). Specifically they showed that longer fluorous carboxylic acid ligands led to the selective and rapid hydrogenation of the C=O group of cinnamaldehyde.

The importance of the inorganic nanoparticle–organic surfactant interface[57] and the use of organic surfactants[58] as soft templates has long been recognized in the synthesis of nanoparticles. For example, crystalline spherical Ag nanoparticles could be prepared in mixed reverse micellar nanoreactors formed from silver and sodium salts of bis(2-ethylhexyl) sulphosuccinate (AOT) in isooctane–water solvent system.[58] Further, $Cu(AOT)_2$–isooctane–water system could be tuned in the presence of various anions to yield nanoparticles of various shapes. Presently these surfactants are predominantly viewed as structure directing adsorbates rather than physical templates.[39]

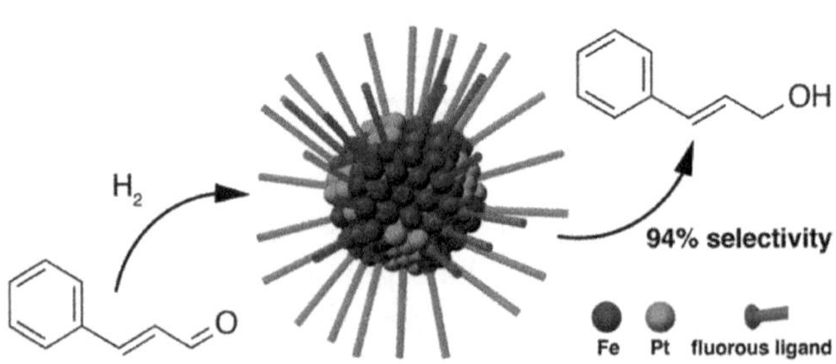

Fig. 2 Schematic representation showing the differential binding affinity of ligand surfactants towards FePt nanoparticles and their application in catalysis. While the Fe atoms are strongly passivated by carboxylic acid surfactants, the surface Pt atoms are only weakly passivated and are available for catalysis. Reprinted with permission from ref. 56, copyright 2015 American Chemical Society.

Cetyltrimethylammonium bromide (CTAB), an ionic surfactant, is known to play a very crucial role in the tunable synthesis of metallic nanorods *via* commonly employed seed-mediated colloidal wet-chemical approaches and other electrochemical and photochemical methods.[59] The surfactants employed in the gold nanoparticle seed formation, for example, sodium citrate or CTAB, can influence their crystallinity and eventually the morphology of the gold nanorods. While the quaternary ammonium cation's chain length is known to impact the aspect ratio of the resulting nanorods, the replacement of the anion *i.e.*, CTAB with cetyltrimethylammonium chloride (CTAC) inhibits the formation of nanorods and spherical nanoparticles are obtained. Murphy and coworkers postulated that the formation of pentatetrahedral twinned nanorods with longer aspect ratios is believed to occur by the blocking of (100) and (110) facets of the growing nanorods by the formation of CTAB bilayers promoting the growth along the (111) direction.[60] Though the detailed mechanistic understanding of the shape-controlled synthesis of nanorods, especially the role of various species involved including CTAB is still incomplete, progress has been made in a number of areas.[59] For example, CTAB's concentration played a critical role in the formation of Pd nanorods, and upon increasing CTAB's concentration the reduction rate of Pd precursor was drastically lowered.[61] In the presence of CTAB, $PdCl_4^{2-}$ is converted to $PdBr_4^{2-}$ which is 10 000 times more stable than the chloride precursor, allowing the preparation of nanorods and triangular platelets under kinetically controlled growth regime.[61] On the other hand, if the reducing agent and seeds are introduced in to a freshly prepared solution containing CTAB and metal precursors (without waiting for embedding of the Pd precursors into the CTAB micelles to allow the exchange of the chloride and bromide) thermodynamically stable Pd nanocubes could be obtained.

Organic amines, oxalate, citric acid, benzoic acid, a solvent like dimethylformamide and carbonmonoxide can act as facet specific capping agents in organic solvents.[14] Their role as facet specific capping agent is not very clear. Oleylamine has been used by a number of groups for the preparation of metal nanowires.[62–64] The available evidence suggests that the mechanism of the reduction of metal salts by oleylamine is dependent on the nature of the metal precursor and could proceed *via* multiple mechanisms.[36] Ligands like oleylamine can modify the metal precursor prior to their reduction. Skrabalak *et al.*[65] showed that irrespective of the nature of Pd metal precursor ($Pd(OAc)_2$, $Pd(acac)_2$, Na_2PdCl_4, palladium hexafluoroacetylacetonate) very similar nanodendrites could be prepared in the presence of oleylamine as reducing agent, solvent and surfactant and attributed it to the incorporation of oleylamine into the coordination sphere of the Pd metal precursors. Additionally they showed that oleylamine complexation could be hindered by other competitive ligands such as trioctylphosphine and PPh_3 and their introduction hinders the reduction, nucleation and growth of nanoparticles in addition to altering their eventual morphology.

Ligand complexation in the context of bimetallic nanoparticles could also invert the core–shell architectures. Though Pd core–Ag shell nanoparticles are expected due to the higher SRP of Pd when compared to Ag, the complexation of Pd in an ammoniacal solution results in the reduction of its SRP leading to the formation of Ag core–Pd shell nanoparticles.[17]

As described above, the ligand surfactants employed as stabilizers or co-stabilizers and the ligands surrounding the metal precursors can play dominant and often intertwined roles in a number of ways during nanoparticle synthesis and our understanding of their influence is evolving over the years.[14,17,36–38,59] In the following sections, we will discuss some select examples from recent literature demonstrating the prowess of coordinating and stabilizing ligands in the controlled synthesis of bi- and multi-metallic nanoparticles.

3 Influence of coordination ligands on metal precursors

3.1 Size effect

LaGrow *et al.* probed the role of metal precursors with various coordination ligands and their oxidation state in the preparation of PtNi and PtCu alloy nanoparticles.[22] While nano-octahedra shaped nanoparticles were obtained in the presence of various metal precursors, there were significant variations in the dimensions and atomic distribution (Fig. 3). The authors observed that the particle size of the resulting nano-octahedra increased in the following order with respect to various precursors: (a) $Pt(acac)_2 < PtCl_2 < PtCl_4 < PtBr_2 < PtBr_4$ (b) $Ni(acac)_2 \cdot 2H_2O < NiCl_2 \cdot 6H_2O < NiBr_2 \cdot 2H_2O$ and (c) $Cu(acac)_2 < CuX_2$. The observed trend clearly showed that both the ligand (acac *vs.* halides) and the oxidation state (Pt(II) *vs.* Pt(IV)) systematically influence the dimensions of the resulting nanoparticles. This could be attributed to the (a) stability of platinum and copper precursors and (b) modification of nickel precursors in the presence of oleylamine and carbon monoxide under reaction conditions. The dimension of the resulting nanoparticles increased nearly linearly with the halide concentration in the case of PtNi nanoalloys with bromide having a larger influence when compared to chloride ions. Interestingly the authors showed that the dimensions of the nanoparticles obtained in the presence of $Pt(acac)_2$ and $NiCl_2$ were comparable to those obtained in the presence of $PtCl_2$ and $Ni(acac)_2$. While the relative ratio of Pt : Ni showed precursor dependent variation with a trend showing a lower Pt : Ni ratio in the presence of halides (with the lowest value for $PtBr_4$), the Pt : Cu ratio did not vary significantly except in the case of $PtBr_4$ which showed a slight decrease. The HAADF-STEM analysis of alloy nanoparticles prepared in the presence of acetylaceonate showed uniform brightness indicating the formation of random alloys (Fig. 3). In contrast, the nanoparticles prepared in the presence of halide precursors showed a bright visible atomic contrast in the edge attributable to the presence of Pt (Fig. 3). Additionally the EDS analysis further corroborated the observation of HAADF, which confirmed the enrichment of Pt close to the surface and Ni or Cu in the center. The authors hypothesized that the segregation in these nanostructures could be attributed to the CO induced surface rearrangement, galvanic exchange reaction and oxidative etching.

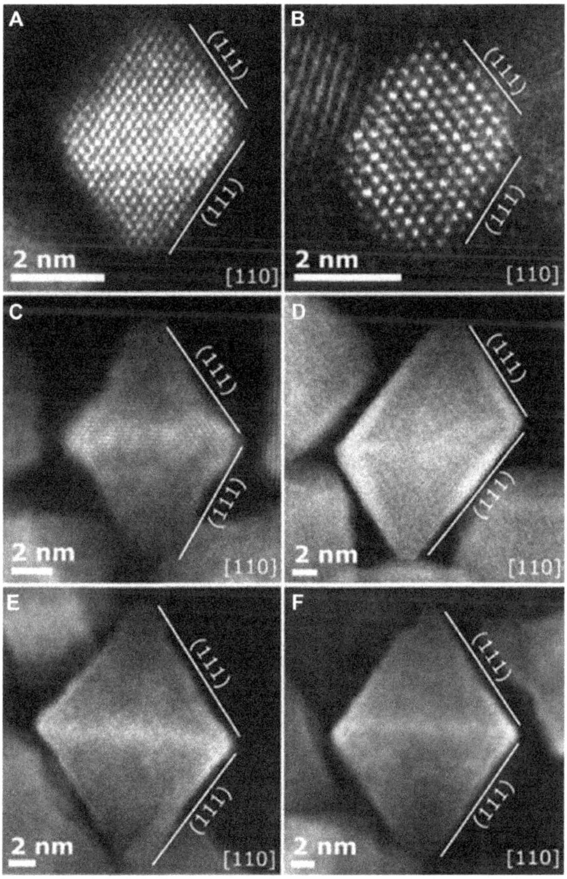

Fig. 3 HAADF-STEM analysis of PtNi and PtCu alloy nanoparticles prepared from various precursors: (A) $Pt(acac)_2$ and $Ni(acac)_2$, (B) $Pt(acac)_2$ and $Cu(acac)_2$ (C) $PtCl_4$ and $NiCl_2 \cdot 6H_2O$, (D) $PtCl_4$ and $CuCl_2$ and (E) $PtBr_4$ and $NiBr_2 \cdot 2H_2O$ and (F) $PtBr_4$ and $CuBr_2$. Reprinted with permission from ref. 22, copyright 2015 American Chemical Society.

Wang and coworkers recently synthesized $[Au_{80}Ag_{30}(C{\equiv}CPh)_{42}Cl_9]^+Cl^-$ (Fig. 4), the largest structurally characterized bimetallic nanoparticle as-of-date, by the reduction of AuC≡CPh and AgCl and showed the importance of halide in its formation.[66] The cationic bimetallic cluster has D_3 symmetry with the metal atoms distributed in a four-shell Russian doll pattern with $Au_6@Au_{35}@Ag_{30}Au_{18}@Au_{21}$. While each of the Au atoms in the periphery are linearly bound to two alkynyl ligands ($21 \times 2 = 42$ alkynyl ligands) to form a monomeric staple motif (PhC≡C–Au–C≡CPh), each of the 9 chloride atoms bridge two Ag atoms in the penultimate shell (Fig. 4). The strong binding of the chloride to Ag atom plays a key role in this synthesis. The major product resulting in the absence of the chloride source was substantially smaller (with a mass peak around 9.3 kDa) instead of 23.56 kDa observed for $[Au_{80}Ag_{30}(C{\equiv}CPh)_{42}Cl_9]^+$.[66] On the other hand, upon $NaBH_4$ reduction of AuC≡CPh and $AgSbF_6$ in the presence of NH_4Cl (as an external chloride source), $Au_{80}Ag_{30}$ could be detected along with an even larger 32 kDa

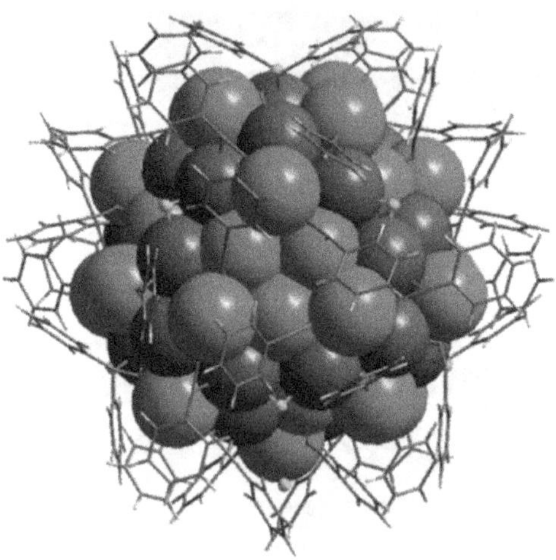

Fig. 4 A schematic representation of $[Au_{80}Ag_{30}(C{\equiv}CPh)_{42}Cl_9]^+Cl^-$. Reprinted with permission from ref. 66, copyright 2016 American Chemical Society.

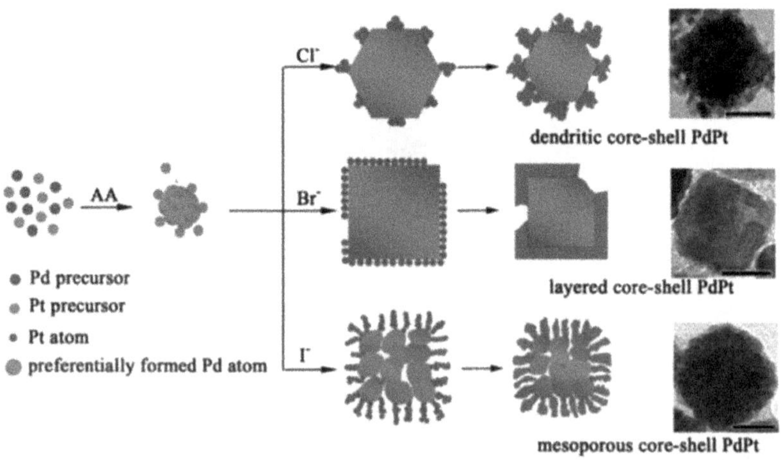

Fig. 5 Schematic representation of the morphology development of PdPt bimetallic nanostructures in the presence of various halides. TEM images (with 20 nm scale bar) are provided for the nanostructures obtained in the presence of 80 mM halide concentration. Reproduced from ref. 23 with permission from The Royal Society of Chemistry.

peak. Notably, chloride could also be entirely replaced with bromide to form $[Au_{80}Ag_{30}(C{\equiv}CPh)_{42}Br_9]^+Br^-$ with AgBr as reactant.[66]

3.2 Shape effect

Hu and coworkers have recently shown the halide dependent preparation of core–shell PdPt nanoparticles with distinct dendritic, layered and mesoporous morphologies (Fig. 5).[23] The chloride metal precursors initially employed in various syntheses are converted to the corresponding halide precursors in the presence of the added halides and reduced with ascorbic acid in the presence of cetylpyridinium chloride. The authors observed the presence of Pd in the core of all these distinct morphologies due to the rapid reduction of Pd precursor over Pt under the reaction conditions.[23] While cuboctahedron Pd nanoparticles (of 40 nm edge dimensions) are obtained initially in the presence of chloride due to oxidative etching, somewhat larger (75 nm) cubic nanoparticles are obtained in the presence of bromide which can protect the Pd surface from etching. On the other hand, aggregates of much smaller Pd nanoparticles were obtained initially in the presence of iodide, presumably due to rapid reduction of iodide precursors. Though initially the dimensions of the Pd cores increased over time, later it decreased due to galvanic exchange reaction with Pt precursors. Halides present in the medium can adsorb on to Pd in the following order: $I^- > Br^- > Cl^-$, and influence the subsequent deposition of Pt on Pd cores. In the presence of chloride ions, the deposition of Pt on Pd cores occurs in island growth mode due to high surface free energy and interatomic bond energy, leading to the rapid formation of dendritic structures.[23] In the presence of bromide ions, the slow generation of Pt atoms leads to its deposition in a layered pattern on the entire surface of Pd core coated with bromide and $PtBr_4$. While strong binding by iodide on Pd cores is also expected to lead to uniform deposition, the outward diffusion of Pd atoms to react with Pt is favored by iodide ions leading to the formation of channels. Interestingly, the dimensions of the morphologically distinct architectures can be altered by varying the halide concentrations.

Very recently, Xia and coworkers reported the role of bromide in modulating the reduction kinetics of Pd and Pt precursors and the ensuing morphology of resulting nanoparticles (Fig. 6).[24] They showed that the polyol reduction of Na_2PdCl_4 and K_2PtCl_4 yielded Pd core–Pt shell nanooctahedra while in the additional presence of KBr the reduction yielded cubic alloy nanoparticles (Fig. 6). They observed that the rate constants for the reduction of $PdCl_4^{2-}$ and $PtCl_4^{2-}$ was significantly different in the absence of KBr with ~92.8% and ~61.8% of the respective metal precursors reduced within the first 2 minutes. The measured rate constant of Pd reduction was 2.7 times that of the Pt reduction. Coupled with the initial concentration of the metal precursors, the reduction rate of Pd was 10 times faster than Pt resulting in the formation of core–shell structures. On the other hand, in the case of the corresponding bromide precursor (produced *in situ* in the presence of KBr) only ~25.4% Pd precursor and ~30.4% Pt precursor were reduced within the first 2 minutes. In the case of bromide precursors, they calculated a relative rate of 2.4. Based on the rate measurements, they concluded that a 4–5 fold difference in reduction rates will yield core–shell architectures.

In the past Yin *et al.* have elegantly employed a combination of halides and organic surface capping agents in the shape controlled synthesis of bimetallic PdPt nanoparticles.[25] They synthesized nanotetrahedrons by the combined use of $Na_2C_2O_4$ and formaldehyde as the (111)-facet selective agent, while nanocubes were obtained in the presence of bromides and iodides.

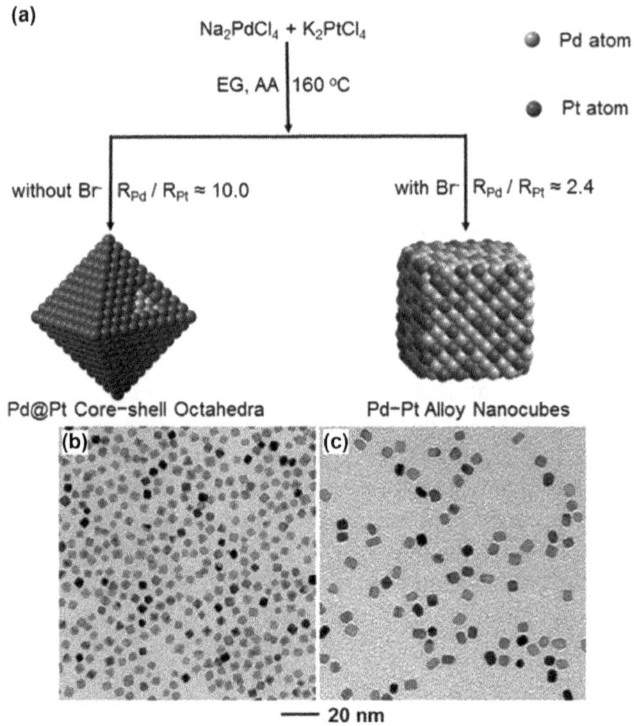

Fig. 6 (a) Schematic representation of the influence of the rate of reduction of Pd and Pt precursors in the absence and presence of bromide resulting in determining the morphology of the resulting nanoparticles, and the TEM images of the resulting core–shell octahedral nanoparticles (b) and nanocubes (c). Reprinted with permission from ref. 24, copyright 2016 American Chemical Society.

4 Influence of linear surfactants

4.1 Size effect

Strasser and coworkers showed the CO assisted solvothermal synthesis of homogeneous bimetallic PtNi nanoparticles of wide ranging composition with a constant dimension of ~4 nm (Fig. 7).[26] The atomic % composition of Ni in the bimetallic nanoparticles could be increased by increasing the reaction temperature from 160 °C to 240 °C in the presence of CO. In the presence of CO irrespective of the temperature employed the size of the nanoparticle remained a constant at 4 nm, while in the absence of CO the size of the nanoparticles increased from 10 to 40 nm.[26] They suggested that the adsorption of CO to PtNi seeds retarded the growth of nanoparticles beyond a critical dimension of 4 nm. Further CO also scavenges Ni *via* complexation to generate $Ni(CO)_4$, limiting the effective concentration of Ni during the synthesis. They attribute the increased Ni composition at higher temperatures to the lability of this complex. The role of CO in the synthesis of metal nanoparticles has been thought of to be limited to a reducing agent and shape directing agent.[36] However the authors[26] showed that raising the CO pressure did not alter the shape of the resulting nanoparticles or accelerate the metal ion reduction in this case.

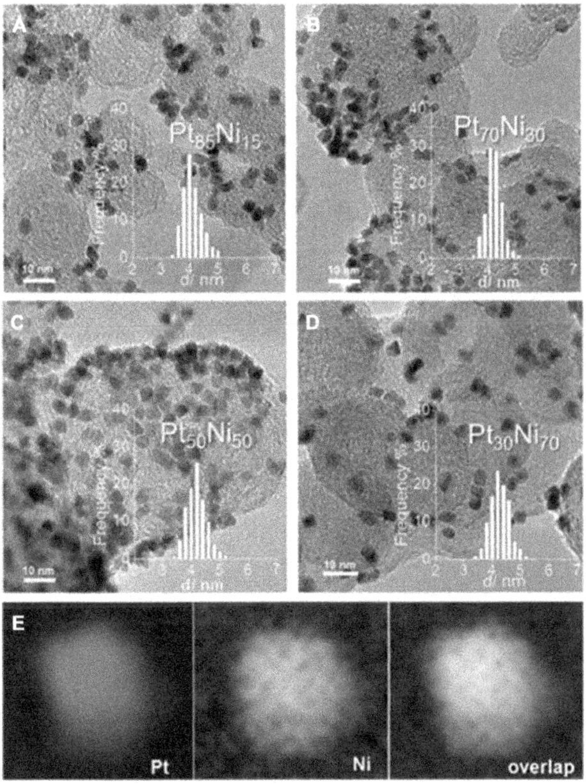

Fig. 7 TEM analyses of PtNi nanoparticles with various compositions prepared at (A) 160 °C (B) 180 °C (C) 200 °C and (D) 240 °C. (E) The EELS elemental map of sample C. Reprinted with permission from ref. 26, copyright 2014 American Chemical Society.

An alternate strategy for the preparation of bimetallic nanoparticles is the spontaneous alloying of pure monometallic nanoparticles, which depends on the nature of the stabilizing ligands.[55] Recently AuPt nanoparticles were prepared by allowing THPO stabilized monometallic Au and Pt nanoparticles to age for over 10 days.[55] The dimensions of the resulting bimetallic nanoparticles (3.1 nm) were in between those of the constituent monometallic nanoparticles (4–6 nm for Au, and 1.8 nm for Pt). Both the stabilizing ligand and the dimensions of the monometallic nanoparticles are crucial in the formation of bimetallic nanoparticles. For example, when strongly binding PEGylated thiol functionalized gold nanoparticles were used, only monometallic nanoparticles were observed even after ageing for 75 days. Further, when citrate stabilized nanoparticles of somewhat larger dimensions (20 nm) were employed instead of THPO stabilized relatively smaller Au nanoparticles, no alloying was observed.[55]

4.2 Shape effect

Li and coworkers showcased the power of organic surfactants such as benzoic acid and aniline in the shape controlled synthesis of Pt_xNi_{1-x} alloy nanoparticles (Fig. 8).[27] While benzoic acid favored the formation of octahedral nanocrystals bound by eight (111) facets (Fig. 8a–d), replacing

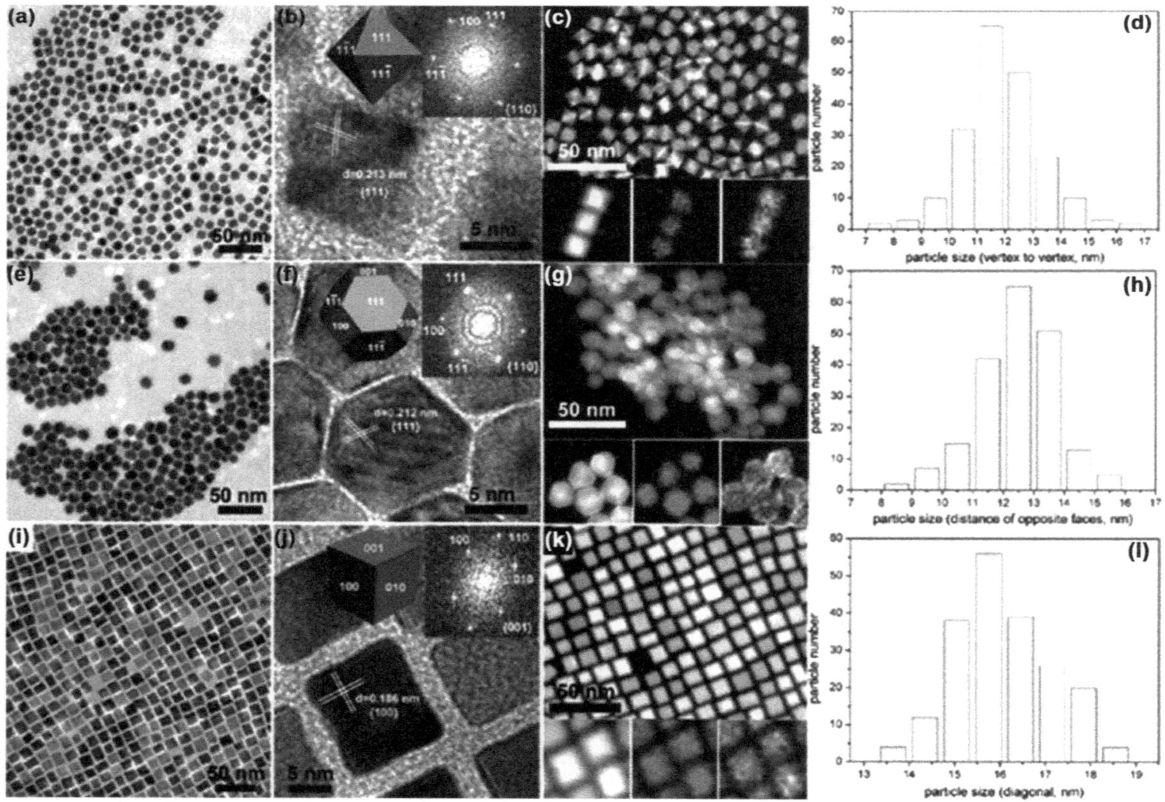

Fig. 8 TEM (a, e, i), HRTEM (b, f, j), HAADF-STEM (c, g, k) and size analyses (d, h, l) of $PtNi_2$ octahedrons (a–d), truncated octahedrons (e–h) and nanocubes (i–l). Reprinted with permission from ref. 27, copyright 2012 American Chemical Society.

it with aniline yielded truncated octahedral shapes bound by both (111) and (100) facets (Fig. 8e–h). On the other hand, the presence of KBr and carbonmonoxide lead to the formation of cubic nanoparticles bound by (100) facets (Fig. 8i–l). Interestingly, the authors showed the stronger adsorption of CO than bromide on (100) facets.

Yang and coworkers reported the synthesis of star shaped Au core–Pt shell decahedra mainly bound by (111) facets in the presence of oleylamine (Fig. 9).[28] The synthesis was accomplished in a one-pot operation where $HAuCl_4$ and H_2PtCl_6 were dissolved in oleylamine and heated in the presence CTAB and trioctylphosphine oxide (TOPO) at 180 °C. The color of the solution turned pink within 1 min. indicating the selective reduction of Au precursor (Fig. 9a), which is expected by the difference in SRP data. HRTEM analysis at this stage revealed the formation of decahedron shaped nanoparticles with 5-fold twins, which is favored under thermodynamically controlled reaction condition. Pt was found deposited on these nanoparticles preferentially at the single crystalline domains instead of the twin boundaries (Fig. 9b). Subsequent facet development was explained in terms of minimization of strain and total surface energies (Fig. 9c). The authors attributed the formation of epitaxial structures *via* Frank–van der Merwe mode growth to the strong adsorption of amine to Pt surfaces leading to dramatic decrease in its surface energy.[28] While oleylamine could be replaced with

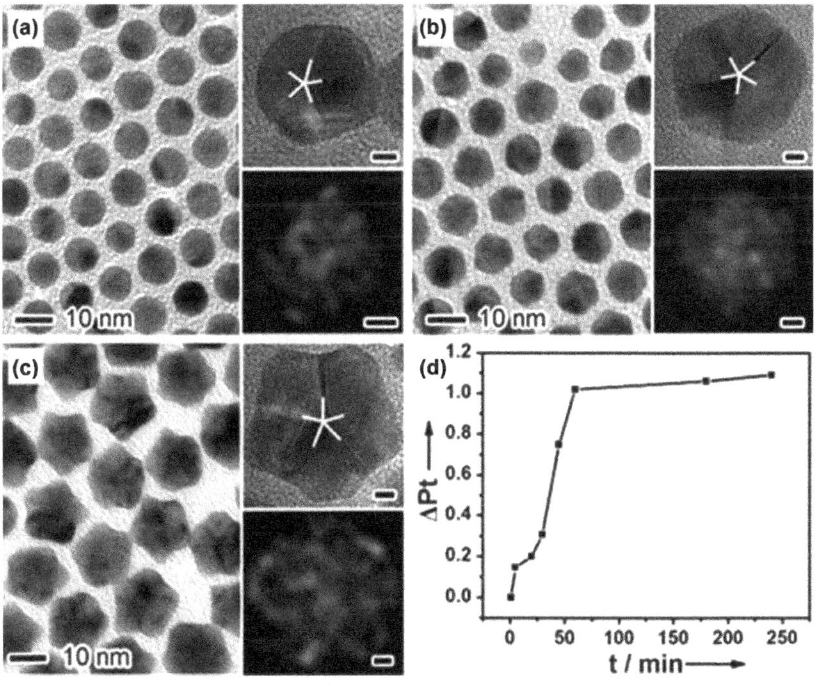

Fig. 9 TEM analyses with insets (scale bar 2 nm) of HRTEM and EDS mapping of Au core–Pt shell star shaped decahedra formation at various time periods: (a) 1 min, (b) 20 min and (c) 60 min. (d) Changes in the mol % of Pt over time. Reprinted with permission from ref. 28, copyright 2015 American Chemical Society.

octadecylamine without morphological changes, replacement with a non-amino compound yielded nanodendrites. CTAB also played a critical role as in its absence only spherical Au core–Pt shell nanoparticles were obtained. The introduction of TOPO in this system ensured the formation of uniform nanostructures.

Recently it was shown that the morphology of CuPd alloy nanoparticles could be carefully controlled to yield tetrahedron, concave tetrahedron, tetrapod and rhombohedral-tetrapod shaped nanoparticles by varying the amount of decylamine.[29] To address the significant difference in the SRP of Pd and Cu precursors, the authors employed decylamine which could coordinate to both these metals and thereby reduce or eliminate the SRP differences. The synthesis was accomplished by heating a mixture of Na_2PdCl_4, $CuCl_2 \cdot 2H_2O$, PVP, glucose in DMF at 110 °C along with varying amounts of decylamine. Decylamine can also selectively stabilize (111) facets allowing the generation of tetrahedral seed nanoparticles, and in its additional growth into other morphologies *via* diffusion-controlled growth.[29] When decylamine was replaced with trioctylamine only smaller spherical nanoparticles were obtained. Notably the morphologically controlled nanostructures obtained in the presence of decylamine were rich in Pd (4–96%). The Pd : Cu ratio of the morphologically distinct nanoparticles could not be altered without altering the morphology of the resulting products. The rate of formation of nanoparticles as monitored by the amount of reduced ions ($[Cu^0]+[Pd^0]$) increased with the amount of decylamine in the reaction medium and this was attributed to the enhanced reducing power of glucose in the presence of amine. Also, it appears that PVP primarily prevents the aggregation of nanoparticles; and does not control the morphology.

The use of octadecyl chain functionalized with two amidoamine headgroups has been shown to lead to bimetallic PdNi nanowires (Fig. 10a and b) and monometallic Pd nanowires (not shown).[30] For the synthesis of bimetallic nanowires, Pd metal ions were phase-transferred from an aqueous solution into toluene phase with the aid of tetraoctylammonium bromide (TOABr) and subsequently treated with a toluene solution of the amidoamine surfactant containing the nickel precursors and stirred at room temperature for 2 h. Subsequently the metal ions in the toluene phase were reduced with an aqueous solution

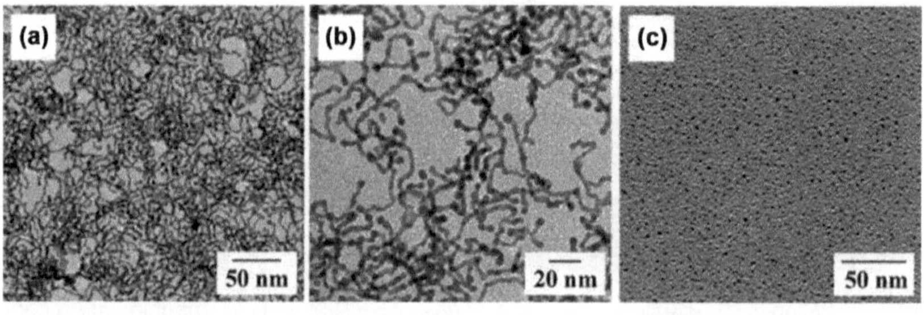

Fig. 10 TEM images of PdNi nanowires (a–b) prepared at room temperature and nanoparticles (c) prepared at 80 °C. Reprinted with permission from ref. 30, copyright 2014 American Chemical Society.

of $NaBH_4$ leading to the formation of nanowires. The authors invoked the diffusion-limited aggregation mechanism in explaining the formation of nanowires under these reaction conditions.[30] The amidoamine surfactant played a crucial role in the formation of the nanowire, as when it was replaced with oleylamine only nanoparticles were observed over a wide range of concentrations. Interestingly, the morphology of the nanowires could be altered by varying the reduction temperature. At 80 °C, with a higher reduction rate the reaction lead to the formation of nanoparticles (Fig. 10c).

PtPdRu nanodendrites could be prepared in a simple operation by the co-reduction of $RuCl_3$, Na_2PdCl_4 and K_2PtCl_4 at room temperature with ascorbic acid in the presence of PVP.[9] When compared to the ratio of the metal precursors employed in the synthesis the dendrites were rich in Pt. Notably, such trimetallic nanodendrites could be etched with HNO_3 to yield hollow dendritic PtPdRu nanocages, which were further enriched in Pt.

Pal and coworkers recently showed the importance of CTAB ligands in mediating the homogeneous and compositionally uniform synthesis of dog-bone or dumbbell shaped core–shell architectures with a gold nanorod core–platinum sandwich–palladium shell (Fig. 11).[31] The positively charged CTA^+ ions present on gold nanorods electrostatically attract Pt ions and also protect the AuPt bimetallic nanoparticles obtained after reduction with ascorbic acid. When they switched CTAB stablilized gold nanorods with citrate stabilized gold nanoparticles, the growth of platinum nanoparticles on gold seeds was hampered. CTA^+ also facilitates the attachment of such bimetallic nanoparticles to negatively charged graphene oxide nanosheets. Subsequently Pd salts are reduced on such bimetallic nanoparticles adsorbed on graphene oxide nanosheets. As silver and bromide ions effectively passivate the side facets of the gold nanorods, the deposition of Pt occurs in the (111) direction resulting in the formation of a dog-bone shaped morphology. Also, somewhat smoother dogbone shaped nanostructures could be fabricated by reducing Pd salt on Au nanorods.

Flower-like, polycrystalline RhPt bimetallic nanoparticles were formed in the presence of CTAC, formaldehyde and citric acid in aqueous medium at 180 °C.[32] The formation of these flower-like structures

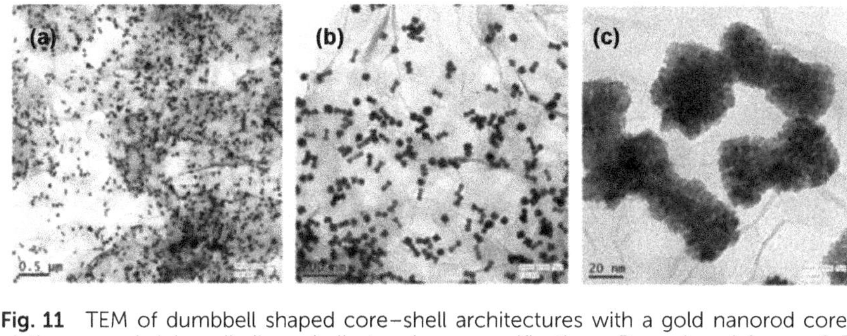

Fig. 11 TEM of dumbbell shaped core–shell architectures with a gold nanorod core–platinum sandwich–palladium shell at various magnifications. Reproduced from ref. 31 with permission from The Royal Society of Chemistry.

required both CTAC and formaldehyde and was suggested to proceed *via* reaction-limited aggregation pathway.[32] The composition of the flower-like nanoparticles could be tuned by varying the precursor concentrations.

Xiong and coworkers recently showed the importance of TOABr in controlling the morphology of trimetallic nanoparticles.[33] They synthesized PtFeCo trimetallic nanoparticles with a Tristar shape and composition of $Pt_{81}Fe_{29}Co_6$ in a one-pot procedure by reducing the metal precursors ($Pt(acac)_2$, $Fe(NO_3)_3$ and $Co(acac)_2$) in 1-octadecene with 1,2-tetradecanediol by heating in the presence of TOABr, oleylamine, oleic acid and 1-dodecanethiol. EDS analysis showed the uniform distribution of the component metals in these anisotropic structures. TOABr played a crucial role in the formation of Tristar structures, as synthesis carried out in its absence under otherwise identical conditions, led to the formation of other random morphologies however with the same composition. The authors attribute the formation of Tristar to the interaction of Pt with TOABr. Remarkably the authors could vary the composition of these Tristar nanoparticles by varying the metal precursor ratio and reported the synthesis of Tristar nanoparticles with the following compositions: $Pt_{81}Fe_{22}Co_5$, $Pt_{81}Fe_{20}Co_7$, and $Pt_{81}Fe_{18}Co_9$. It is worth noting that a very similar synthetic protocol was earlier adopted for generating high-quality, cubic, bimetallic PtCu nanoparticles with tunable compositions, where TOABr along with oleylamine played a crucial part.[67]

4.3 Other influences

A supramolecular approach was developed for the reliable synthesis of core–shell nanoparticles (Fig. 12).[68] In this seed mediated approach, a functionalized ligand capable of binding metal ions was employed to circumvent the formation of monometallic nanoparticles as side product. Au and Pd nanoparticles were synthesized by a Brust–Schiffrin type two phase-reduction in the presence of an imidazole ligand functionalized with an amide group. To these seed nanoparticles, the complimentary metal salt (*i.e.*, Au salt to Pd nanoparticles and *vice versa*) were added as their tetrabutylammonium salts, stirred for 15 minutes and subsequently reduced with sodium borohydride, leading to the preparation of core–shell nanoparticles as determined by HAADF-STEM analysis. When the imidazole ligand bearing the amide functionality was replaced with *N*-dodecylimidazole, agglomerated bulk metal was observed during the second stage reduction highlighting the importance of the amide functionality and its ability to bind to the chlorometallate anion.

A related strategy was adopted for the synthesis of core–shell multimetallic architecture containing an iron oxide core and a gold shell for use in the simultaneous magnetic resonance and photoacoustic hybrid imaging of cells.[12] In this approach, the aldehyde groups present in the dextran ligands on iron oxide nanoparticles reduced the diammine-silver(I) nitrate leading to the formation of silver nanoparticles on iron oxide surfaces. Such iron oxide templated silver nanoparticles acted as seeds for the subsequent deposition of gold in the presence of citrate.

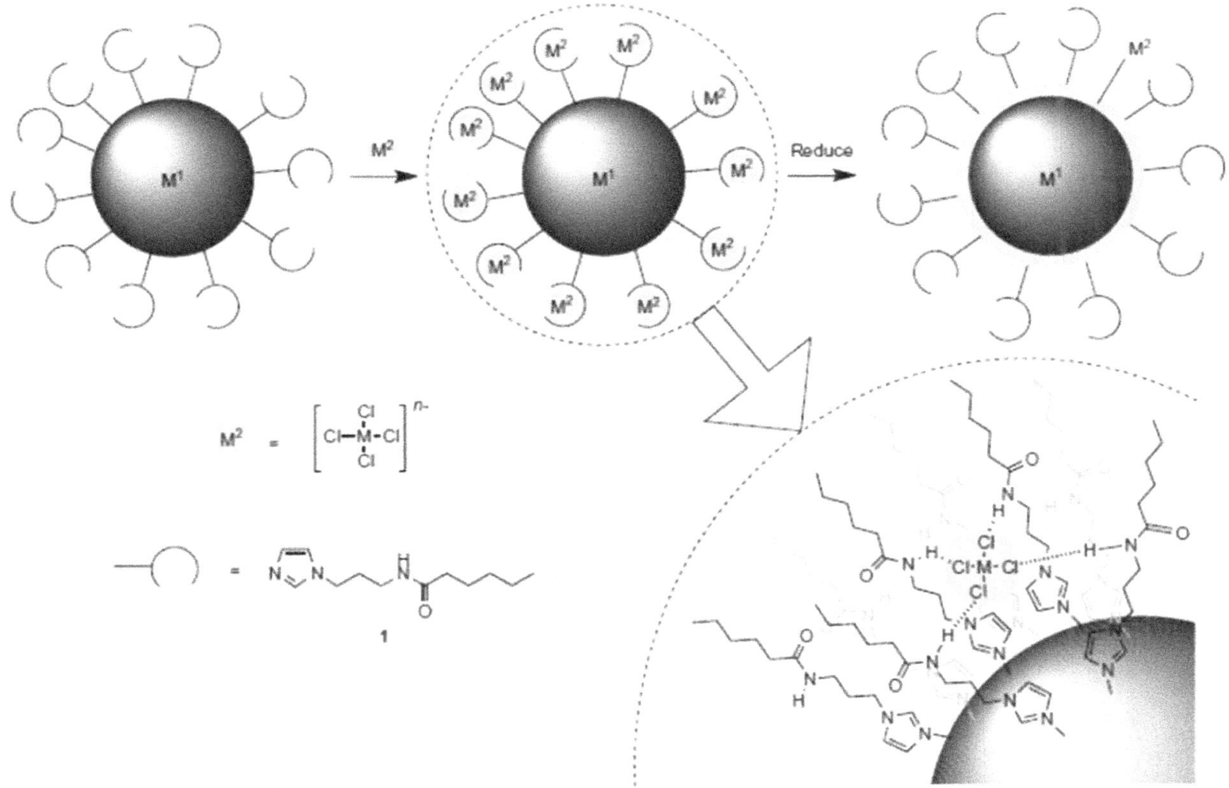

Fig. 12 A supramolecular approach for the seed-mediated synthesis of core–shell bimetallic nanoparticles. Reprinted by permission from Macmillan Publishers Ltd: Nature Chemistry, ref. 68, copyright 2011.

Attempts to directly deposit gold (without silver) resulted in the formation of much larger, unstable dispersions.

Jose-Yacaman and coworkers reported the synthesis of spherical trimetallic nanoparticles comprising of a AuCu core–Pt shell (Fig. 13) in a one-pot procedure in the presence of oleylamine, which is addition to be being a ligand surfactant can also act as a reducing agent and solvent.[69] In this report, gold salt was initially reduced and the resulting gold nanoparticles (9.0 ± 1.0 nm) were converted to AuCu alloy nanoparticles (13.0 ± 1.0 nm) by the addition and reduction of a copper salt. Subsequently platinum salt was added to the alloy nanoparticles and reduced yielding core–shell trimetallic nanoparticles (15.0 ± 1.0 nm).[69] The presence of AuCu core–Pt shell was confirmed by both EDS and HAADF analysis (Fig. 13).

Seed-mediated co-reduction strategy was recently adapted for the synthesis of trimetallic nanoparticles.[35] On Ag nanocube seeds, $HAuCl_4$ and H_2PdCl_4 were co-reduced with ascorbic acid a weak reductant, in the presence of CTAC. Both Au and Pd salts can undergo galvanic exchange with the Ag seed under these conditions.[19] SEM, TEM and STEM-EDS

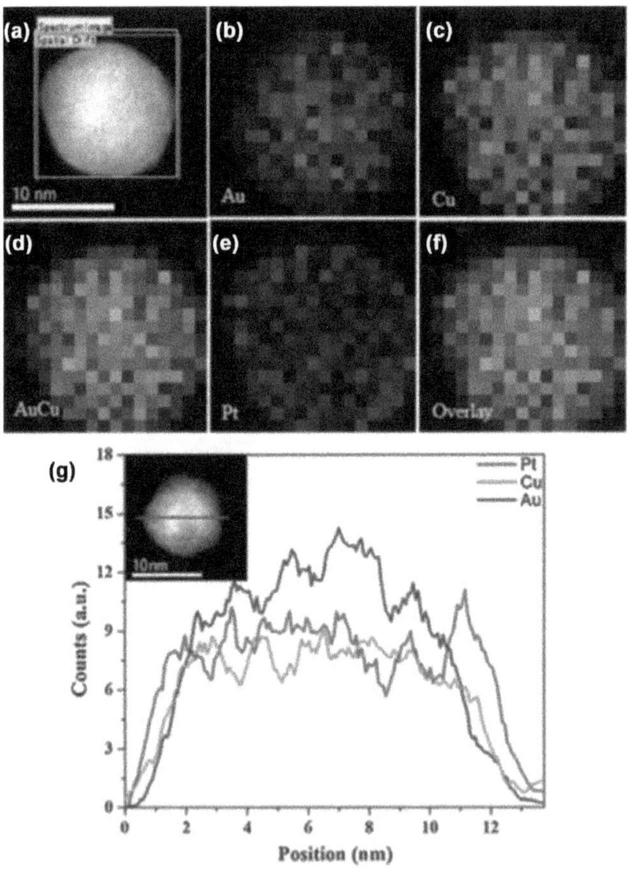

Fig. 13 HAADF-STEM (a), EDS (b–f) and elemental line analysis (g) of AuCu core–Pt shell nanoparticles. Reproduced from ref. 69 with permission from The Royal Society of Chemistry.

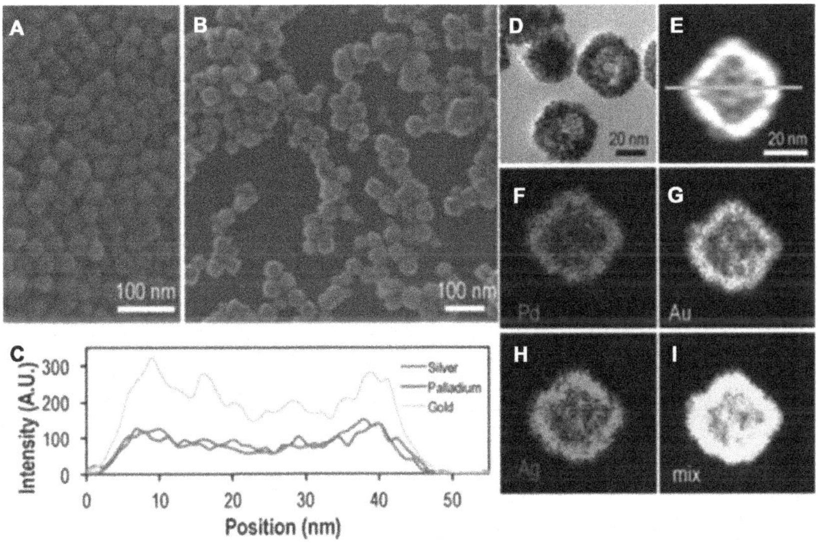

Fig. 14 SEM (A, B) and TEM (D) of silver nanocube seeds (A), and the seed-mediated coreduction product (B, D). F–I shows the EDS elemental mapping and associated EDS linescan (C and E). Reproduced from ref. 35 with permission from The Royal Society of Chemistry.

analyses reveals the formation of hollow trimetallic nanoparticles (Fig. 14).[35] To delineate the importance of galvanic exchange and seed-mediated reduction processes, the authors carried out the reaction in the absence of ascorbic acid. Though the galvanic exchange proceeded to yield hollow nanocubes significant fragmentation was also observed. The surfactant present during the galvanic exchange can play a major role in the integrity of the final product.[19] Interestingly, by increasing the pH of the reaction medium the reduction power of the ascorbic acid could be increased and thereby the galvanic exchange process could be substantially suppressed leading to the formation of Ag core–AuPd shell trimetallic nanoparticles.[35]

5 Influence of macrocyclic surfactants

The use of surfactants based on well-established macrocyclic supramolecular building blocks for the stabilization and functionalization of nanoparticles is appealing as they can potentially lead to enhanced stability, permit supramolecular interactions, and be advantageous in catalytic applications.[70–73] Multidentate surfactants based on macrocyclic molecules such as cyclodextrin, calixarene, resorcinarene, pillararene, and cucurbituril have been used in the synthesis, stabilization, self-assembly and functionalization of metal nanoparticles.[70–77] These macrocycles in their native and modified forms have been used for the synthesis of nominally spherical nanoparticles.

5.1 Size effect

The cyclodextrin surfactant containing secondary hydroxyl groups can reduce the metal ions under appropriate conditions in the absence of

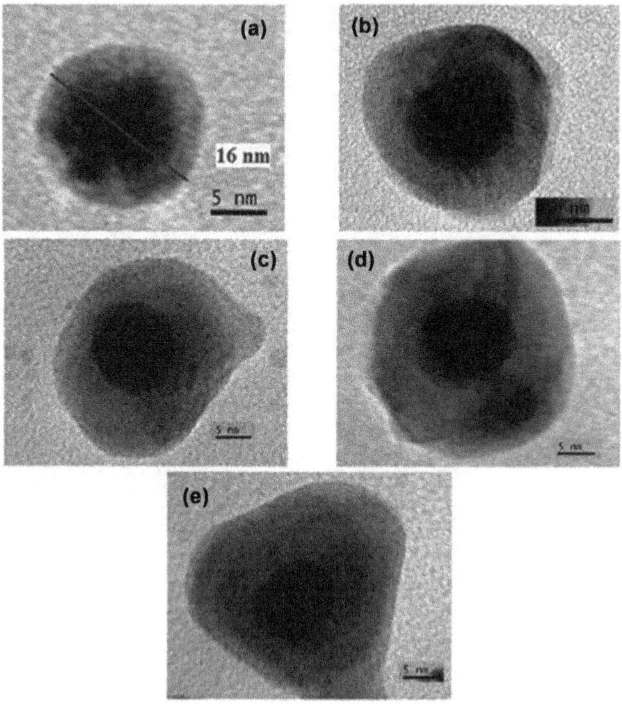

Fig. 15 HRTEM images Au core–Ag shell bimetallic nanoparticles with varying shell thickness. Reprinted with permission from ref. 78, copyright 2011 American Chemical Society.

other reducing agents. Monometallic gold and bimetallic Au core–Ag shell nanoparticles could be obtained in the presence of unmodified β-cyclodextrin within 15 minutes by merely heating the reaction mixture at 90 °C under basic conditions.[78] Further, by varying the relative proportion of $HAuCl_4$ (employed to generate β-cyclodextrin protected gold nanoparticle cores) and $AgNO_3$ (subsequently added to generate the shell), the relative thickness of the Au core and the Ag shell in these bimetallic nanoparticles could be altered (Fig. 15).[78] Recently, it was shown that mono- and bi-metallic nanoparticles could be synthesized even at room temperature in the presence of γ-cyclodextrin.[79]

5.2 Shape effect

Recently,[34] our group showed that a multidentate resorcinarene amine surfactant (Fig. 16) can lead to the formation of V-shaped anisotropic PdPt bimetallic nanoparticles under Brust–Schiffrin conditions. Briefly, a mixture of TOA+ stabilized $PtCl_6^{2-}$ and $PdCl_4^{2-}$ in chloroform was reduced with excess aqueous sodium borohydride at 0 °C in the presence of resorcinarene amine. TEM analysis (Fig. 16a) of the purified product obtained after a simple precipitation (from a mixture of chloroform–isopropanol)–redispersion (chloroform) revealed the dominant formation of nanoparticles with a V-shape or a variant on the V-shape along with a few spherical nanoparticles. While the HRTEM d-spacing

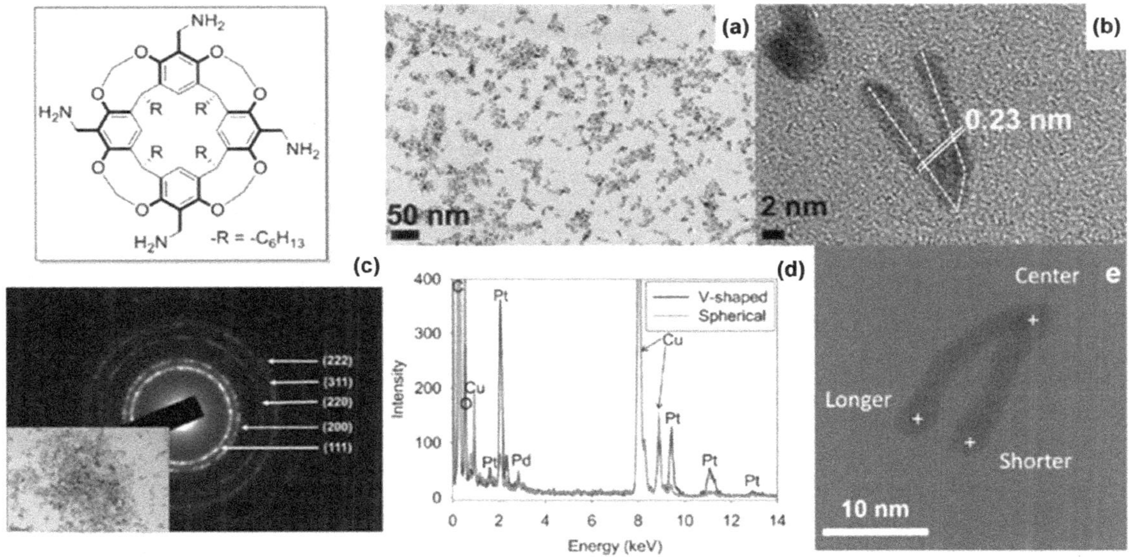

Fig. 16 (a–b) TEM, (c) SAED, and (e) STEM spot analysis of PdPt bimetallic nanoparticles prepared with resorcinarene amine at 0 °C for 2 h under Brust–Schiffrin reaction conditions. (d) EDS analysis of V-shaped and spherical nanoparticles formed in as prepared samples. Adapted from ref. 34 with permission from The Royal Society of Chemistry.

(Fig. 16b) of a V-shaped nanoparticle was consistent with (111) planes of Pt and Pd, its selected area electron diffraction (Fig. 16c) pattern could be indexed to (111), (200), (220), (311) and (222) planes of Pd or Pt. Single particle EDS analysis (Fig. 16d) showed that the V-shaped nanoparticles were significantly rich in Pt (Pt : Pd ratio of 8.7 ± 2.9) when compared to the 1 : 1 ratio of Pt and Pd precursors employed in this synthesis, while the spherical particles showed a Pt : Pd ratio of 0.75 (Fig. 16d). STEM spot analysis (Fig. 16e) of 5 different particles showed that the Pt : Pd ratio at the center of the V-shaped nanoparticle (3.9 ± 0.8) was consistently lower than the ratios observed at the end of the relatively longer (5.6 ± 3.2) or shorter (6.9 ± 2.3) arms. The spontaneous formation of V-shaped and spherical nanoparticles with such significant difference in their composition is unprecedented in the light of work by other groups[80–82] on the synthesis of PdPt nanoparticles using a range of methods who obtained nanoparticles with a composition consistent with the precursor concentrations. Also the formation of such bipodal anisotropic structures under these conditions is remarkable, as often Brust–Schiffrin synthesis[83] yields nominally spherical small nanoparticles[84] and very rarely other morphologies (for example, nanowire)[30] have been reported.

Our group probed certain key experimental variables involved in the formation of V-shaped nanoparticles in detail.[34] Within 5 minutes after the addition of sodium borohydride, both V-shaped and spherical nanoparticles were observed (Fig. 17a). Even after 2 hours of reduction spherical nanoparticles were still present. Also, there was a discernable difference in the dimension of the spherical nanoparticles (2.0 ± 0.4 nm) formed and the cross-sectional dimension of the V-shaped nanoparticle (2.6 ± 0.4 nm). We believe that both anisotropic V-shaped and spherical nanoparticles are formed concurrently from homogeneous nucleation events.[85–87] V-shaped nanoparticles could also be obtained at room

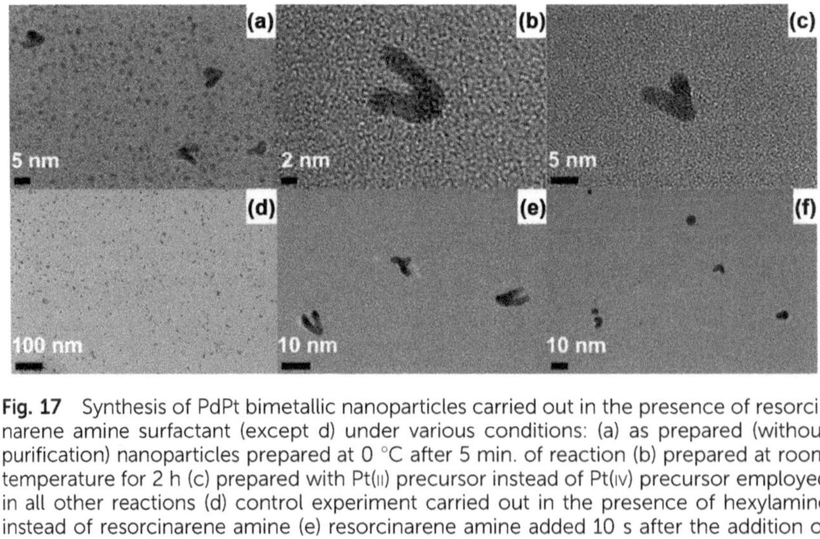

Fig. 17 Synthesis of PdPt bimetallic nanoparticles carried out in the presence of resorcinarene amine surfactant (except d) under various conditions: (a) as prepared (without purification) nanoparticles prepared at 0 °C after 5 min. of reaction (b) prepared at room temperature for 2 h (c) prepared with Pt(II) precursor instead of Pt(IV) precursor employed in all other reactions (d) control experiment carried out in the presence of hexylamine instead of resorcinarene amine (e) resorcinarene amine added 10 s after the addition of sodium borohydride and (f) resorcinarene amine added 5 min after the addition of sodium borohydride. Adapted from ref. 34 with permission from The Royal Society of Chemistry.

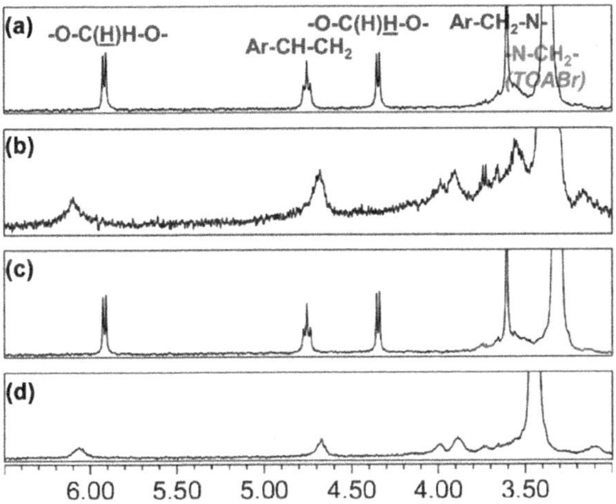

Fig. 18 ^{1}H-NMR spectra of a mixture of TOABr and resorcinarene amine in the presence of various phase-transferred metal salts: (a) control experiment in the absence of metal salts (b) in the presence of both Pt and Pd salts (c) Pt salt and (d) Pd salts. The broadening and shift of the proton peaks is observed only in the presence of Pd salt. Reproduced from ref. 34 with permission from The Royal Society of Chemistry.

temperature (Fig. 17b) albeit with a reduced Pt : Pd ratio of 5.6 ± 1.1. Switching the Pt(IV) precursor to a Pt(II) precursor by replacing H_2PtCl_6 with K_2PtCl_4 also did not impede the formation of V-shaped nanoparticles (Fig. 17c). These nanoparticles had a slightly higher Pt content (Pt : Pd ratio of 13.9 ± 2.6). On the other hand, upon replacing resorcinarene amine with hexylamine surfactant a mixture of dominantly spherical and slightly elongated nanoparticles (with a Pt : Pd ratio of $\sim$1) were obtained (Fig. 17d). Our experiments showed that resorcinarene amine must be present prior to or almost immediately after the addition of sodium borohydride for the formation of V-shaped nanoparticles. When resorcinarene amine was added 10 seconds after the addition of sodium borohydride, somewhat similar crystalline, V-shaped nanoparticles (Fig. 17e) however with a lower Pt : Pd ratio of 5.5 ± 0.6 were obtained. On the other hand, with a 5 minute delay V-shaped or Pt-rich nanoparticles were not obtained. Proton NMR analysis clearly revealed the binding of the palladium salt to resorcinarene amine (preferentially over platinum) prior to their reduction (Fig. 18). These experiments unambiguously revealed the crucial role played by the multidentate resorcinarene amine surfactant in determining the shape, composition and crystallinity of PdPt bimetallic nanoparticles.

6 Conclusion

In this chapter we have showcased the influence of ligands surrounding the metal precursor, and ligand surfactants in dictating the size, shape and compositional control over bi- and multi-metallic nanoparticles. The anionic ligands coordinating to the metal precursor can significantly alter its reactivity and stability, and additionally can play a critical role in determining the shape of the resulting nanoparticles. Organic surfactants routinely enable the synthesis of a variety of anisotropic nanoparticles with potential applications in a number of areas. Resorcinarene amine, a multidentate macrocyclic surfactant, recently facilitated the formation of novel V-shaped alloy nanoparticles with additional control over the composition by a simple co-reduction.[34] Though the impact of organic surfactants stabilizing the nanoparticles on the size, shape and composition of the resulting alloy nanoparticles is clear, the mechanistic understanding of the underlying process is yet to be completely elucidated. One of the current challenges is the *in situ* modification of some of the reactants or reagents during the course of the nanoparticle synthesis, which inadvertently changes the course of the reaction. With a better understanding of such processes it will be possible to design nanoparticle synthesis by tailoring the reactivity of various species and the eventual shape of the resulting nanoparticle in a rational manner.

References

1 J. Jellinek, *Faraday Discuss.*, 2008, **138**, 11.
2 C. F. Cheng and C. M. Schwitter, *Am. J. Archaeol.*, 1957, **61**, 351.
3 F. Grazzi, F. Civita, A. Williams, A. Scherillo, E. Barzagli, L. Bartoli, D. Edge and M. Zoppi, *Anal. Bioanal. Chem.*, 2011, **400**, 1493.
4 R. Ferrando, J. Jellinek and R. L. Johnston, *Chem. Rev.*, 2008, **108**, 845.
5 P.-C. Chen, X. Liu, J. L. Hedrick, Z. Xie, S. Wang, Q.-Y. Lin, M. C. Hersam, V. P. Dravid and C. A. Mirkin, *Science*, 2016, **352**, 1565.
6 A. Zaleska-Medynska, M. Marchelek, M. Diak and E. Grabowska, *Adv. Colloid Interface Sci.*, 2016, **229**, 80.
7 K. D. Gilroy, A. Ruditskiy, H.-C. Peng, D. Qin and Y. Xia, *Chem. Rev.*, 2016, **116**, 10414.
8 K. An and G. A. Somorjai, *Catal. Lett.*, 2015, **145**, 233.
9 K. Eid, H. Wang, V. Malgras, Z. A. Alothman, Y. Yamauchi and L. Wang, *J. Phys. Chem. C*, 2015, **119**, 19947.
10 N. S. Porter, H. Wu, Z. Quan and J. Fang, *Acc. Chem. Res.*, 2013, **46**, 1867.
11 L. Tian, L. Liu, Y. Li, Q. Wei and W. Cao, *Sci. Rep.*, 2016, **6**, 30849.
12 A. A. Bogdanov, A. J. Dixon, S. Gupta, L. Zhang, S. Zheng, M. S. Shazeeb, S. Zhang and A. L. Klibanov, *Bioconjugate Chem.*, 2016, **27**, 383.
13 B. N. Wanjala, J. Luo, B. Fang, D. Mott and C.-J. Zhong, *J. Mater. Chem.*, 2011, **21**, 4012.
14 J. Gu, Y.-W. Zhang and F. Tao, *Chem. Soc. Rev.*, 2012, **41**, 8050.
15 N. Toshima and T. Yonezawa, *New J. Chem.*, 1998, **22**, 1179.
16 M.-L. Wu, D.-H. Chen and T.-C. Huang, *Langmuir*, 2001, **17**, 3877.
17 D. V. Goia and E. Matijevic, *New J. Chem.*, 1998, **22**, 1203.
18 C. Wang, H. Yin, R. Chan, S. Peng, S. Dai and S. Sun, *Chem. Mater.*, 2009, **21**, 433.
19 Y.-S. Shon, G. B. Dawson, M. Porter and R. W. Murray, *Langmuir*, 2002, **18**, 3880.
20 X. Gong, Y. Yang and S. Huang, *J. Phys. Chem. C*, 2010, **114**, 18073.
21 R. G. Weiner, C. J. DeSantis, M. B. T. Cardoso and S. E. Skrabalak, *ACS Nano*, 2014, **8**, 8625.
22 A. P. LaGrow, K. R. Knudsen, N. M. AlYami, D. H. Anjum and O. M. Bakr, *Chem. Mater.*, 2015, **27**, 4134.

23 J. Zhang, L. Wan, L. Liu, Y. Deng, C. Zhong and W. Hu, *Nanoscale*, 2016, **8**, 3962.
24 M. Zhou, H. Wang, M. Vara, Z. D. Hood, M. Luo, T.-H. Yang, S. Bao, M. Chi, P. Xiao, Y. Zhang and Y. Xia, *J. Am. Chem. Soc.*, 2016, **138**, 12263.
25 A.-X. Yin, X.-Q. Min, Y.-W. Zhang and C.-H. Yan, *J. Am. Chem. Soc.*, 2011, **133**, 3816.
26 C. Cui, L. Gan, M. Neumann, M. Heggen, B. Roldan Cuenya and P. Strasser, *J. Am. Chem. Soc.*, 2014, **136**, 4813.
27 Y. Wu, S. Cai, D. Wang, W. He and Y. Li, *J. Am. Chem. Soc.*, 2012, **134**, 8975.
28 T. Bian, H. Zhang, Y. Jiang, C. Jin, J. Wu, H. Yang and D. Yang, *Nano Lett.*, 2015, **15**, 7808.
29 L. Zhang, H. Su, M. Sun, Y. Wang, W. Wu, T. Yu and J. Zeng, *Nano Res.*, 2015, **8**, 2415.
30 Y. Imura, K. Tsujimoto, C. Morita and T. Kawai, *Langmuir*, 2014, **30**, 5026.
31 S. Dutta, C. Ray, A. K. Sasmal, Y. Negishi and T. Pal, *J. Mater. Chem. A*, 2016, **4**, 3765.
32 Q. Yuan, D.-B. Huang, H.-H. Wang and Z.-Y. Zhou, *Langmuir*, 2014, **30**, 5711.
33 N. Du, C. Wang, X. Wang, Y. Lin, J. Jiang and Y. Xiong, *Adv. Mater.*, 2016, **28**, 2077.
34 S. Han, V. P. Sheela, W. Cao and R. Balasubramanian, *RSC Adv.*, 2013, **3**, 8551.
35 R. G. Weiner, A. F. Smith and S. E. Skrabalak, *Chem. Commun.*, 2015, **51**, 8872.
36 N. Ortiz and S. E. Skrabalak, *Langmuir*, 2014, **30**, 6649.
37 M. L. Personick and C. A. Mirkin, *J. Am. Chem. Soc.*, 2013, **135**, 18238.
38 A. J. Biacchi and R. E. Schaak, *ACS Nano*, 2011, **5**, 8089.
39 A. R. Tao, S. Habas and P. Yang, *Small*, 2008, **4**, 310.
40 S. E. Lohse, N. D. Burrows, L. Scarabelli, L. M. Liz-Marzán and C. J. Murphy, *Chem. Mater.*, 2014, **26**, 34.
41 H. Zhang, M. Jin, J. Wang, W. Li, P. H. C. Camargo, M. J. Kim, D. Yang, Z. Xie and Y. Xia, *J. Am. Chem. Soc.*, 2011, **133**, 6078.
42 Y. Xiong, J. M. McLellan, Y. Yin and Y. Xia, *Angew. Chem., Int. Ed.*, 2007, **46**, 790.
43 A. Roucoux, J. Schulz and H. Patin, *Chem. Rev.*, 2002, **102**, 3757.
44 A. C. Templeton, M. P. Wuelfing and R. W. Murray, *Acc. Chem. Res.*, 2000, **33**, 27.
45 J. C. Love, L. A. Estroff, J. K. Kriebel, R. G. Nuzzo and G. M. Whitesides, *Chem. Rev.*, 2005, **105**, 1103.
46 A. Ulman, *Chem. Rev.*, 1996, **96**, 1533.
47 L. S. Ott and R. G. Finke, *Coord. Chem. Rev.*, 2007, **251**, 1075.
48 R. W. J. Scott, O. M. Wilson and R. M. Crooks, *J. Phys. Chem. B*, 2005, **109**, 692.
49 L. M. Bronstein and Z. B. Shifrina, *Chem. Rev.*, 2011, **111**, 5301.
50 S. E. Eklund and D. E. Cliffel, *Langmuir*, 2004, **20**, 6012.
51 C. A. Stowell and B. A. Korgel, *Nano Lett.*, 2005, **5**, 1203.
52 K. G. Thomas, J. Zajicek and P. V. Kamat, *Langmuir*, 2002, **18**, 3722.
53 S. K. Ghosh, S. Nath, S. Kundu, K. Esumi and T. Pal, *J. Phys. Chem. B*, 2004, **108**, 13963.
54 S. T. Marshall, M. O'Brien, B. Oetter, A. Corpuz, R. M. Richards, D. K. Schwartz and J. W. Medlin, *Nat. Mater.*, 2010, **9**, 853.
55 L. Uson, V. Sebastian, A. Mayoral, J. L. Hueso, A. Eguizabal, M. Arruebo and J. Santamaria, *Nanoscale*, 2015, **7**, 10152.
56 K. B. Vu, K. V. Bukhryakov, D. H. Anjum and V. O. Rodionov, *ACS Catal.*, 2015, **5**, 2529.

57 Y. Yin and A. P. Alivisatos, *Nature*, 2005, **437**, 664.
58 M.-P. Pileni, *Nat. Mater.*, 2003, **2**, 145.
59 S. E. Lohse and C. J. Murphy, *Chem. Mater.*, 2013, **25**, 1250.
60 C. J. Murphy, L. B. Thompson, A. M. Alkilany, P. N. Sisco, S. P. Boulos, S. T. Sivapalan, J. A. Yang, D. J. Chernak and J. Huang, *J. Phys. Lett.*, 2010, **1**, 2867.
61 G. Berhault, M. Bausach, L. Bisson, L. Becerra, C. Thomazeau and D. Uzio, *J. Phys. Chem. C*, 2007, **111**, 5915.
62 A. Halder and N. Ravishankar, *Adv. Mater.*, 2007, **19**, 1854.
63 Z. Huo, C.-k. Tsung, W. Huang, X. Zhang and P. Yang, *Nano Lett.*, 2008, **8**, 2041.
64 X. Lu, M. S. Yavuz, H.-Y. Tuan, B. A. Korgel and Y. Xia, *J. Am. Chem. Soc.*, 2008, **130**, 8900.
65 N. Ortiz and S. E. Skrabalak, *Angew. Chem., Int. Ed.*, 2012, **51**, 11757.
66 J.-L. Zeng, Z.-J. Guan, Y. Du, Z.-A. Nan, Y.-M. Lin and Q.-M. Wang, *J. Am. Chem. Soc.*, 2016, **138**, 7848.
67 D. Xu, S. Bliznakov, Z. Liu, J. Fang and N. Dimitrov, *Angew. Chem., Int. Ed.*, 2010, **49**, 1282.
68 C. J. Serpell, J. Cookson, D. Ozkaya and P. D. Beer, *Nat. Chem.*, 2011, **3**, 478.
69 S. Khanal, N. Bhattarai, D. McMaster, D. Bahena, J. J. Velazquez-Salazar and M. Jose-Yacaman, *Phys. Chem. Chem. Phys.*, 2014, **16**, 16278.
70 V. Montes-García, J. Pérez-Juste, I. Pastoriza-Santos and L. M. Liz-Marzán, *Chem. – Eur. J.*, 2014, **20**, 10874.
71 N. Evans, H. Rahman, J. Davis and P. Beer, *Anal. Bioanal. Chem.*, 2012, **402**, 1739.
72 H. Li and Y.-W. Yang, *Chin. Chem. Lett.*, 2013, **24**, 545.
73 A. Wei, *Chem. Commun.*, 2006, 1581.
74 J. Alvarez, J. Liu, E. Roman and A. E. Kaifer, *Chem. Commun.*, 2000, 1151.
75 L. Strimbu, J. Liu and A. E. Kaifer, *Langmuir*, 2003, **19**, 483.
76 Y. Sun, Y. Yao, C.-G. Yan, Y. Han and M. Shen, *ACS Nano*, 2010, **4**, 2129.
77 V. Huc and K. Pelzer, *J. Colloid Interface Sci.*, 2008, **318**, 1.
78 S. Pande, J. Chowdhury and T. Pal, *J. Phys. Chem. C*, 2011, **115**, 10497.
79 V. I. Bhoi, S. Kumar and C. N. Murthy, *New J. Chem.*, 2016, **40**, 1396.
80 N. Toshima, M. Harada, T. Yonezawa, K. Kushihashi and K. Asakura, *J. Phys. Chem.*, 1991, **95**, 7448.
81 H. Ye and R. M. Crooks, *J. Am. Chem. Soc.*, 2007, **129**, 3627–3633.
82 Y. Liu, M. Chi, V. Mazumder, K. L. More, S. Soled, J. D. Henao and S. Sun, *Chem. Mater.*, 2011, **23**, 4199.
83 M. Brust, M. Walker, D. Bethell, D. J. Schiffrin and R. Whyman, *J. Chem. Soc., Chem. Commun.*, 1994, 801.
84 M. J. Hostetler, J. E. Wingate, C.-J. Zhong, J. E. Harris, R. W. Vachet, M. R. Clark, J. D. Londono, S. J. Green, J. J. Stokes, G. D. Wignall, G. L. Glish, M. D. Porter, N. D. Evans and R. W. Murray, *Langmuir*, 1998, **14**, 17.
85 B. Lim, M. J. Jiang, T. Yu, P. H. C. Camargo and Y. N. Xia, *Nano Res.*, 2010, **3**, 69.
86 B. Lim and Y. Xia, *Angew. Chem., Int. Ed.*, 2011, **50**, 76.
87 B. Lim, M. Jiang, P. H. C. Camargo, E. C. Cho, J. Tao, X. Lu, Y. Zhu and Y. Xia, *Science*, 2009, **324**, 1302.

Recent highlights in advanced transmission electron microscopy techniques: applications to nanomaterials

Francis Leonard Deepak*[a] and Gilberto Casillas[b]
DOI: 10.1039/9781782620358-00029

In this chapter, recent advances in scanning/transmission electron microscopic techniques and their applications to nanomaterials are highlighted. Aberration corrected electron microscopy is now routinely employed to study carbon nanomaterials, nanoparticles, 2D materials, alloys and heterostructures, *etc.* However, there have been several significant achievements in employing this technique towards the study of the various nanostructures. Aberration corrected TEM/STEM imaging tends to be very successful in identifying new nanoalloys, enables understanding of the growth mechanisms as well as the behaviour of supported atoms/clusters towards catalytic reactions, *etc.* HAADF-STEM imaging in combination with spectroscopy is an essential tool for the characterization of new 2D materials, such as borophenes and phosphorenes. At the same time, it provides insights into the structure of hetero-interfaces as well as enables identifying defects and new phases associated with these materials. It is now possible to go down in accelerating voltages upto 15 kV in both TEM/STEM modes and image specimens with minimal damage. It is foreseen, that these new developments in advanced electron microscopic techniques will provide significant possibilities for unravelling new phenomena, as well as explore novel properties of nanomaterials enabling new applications.

1 Introduction to aberration corrected S/TEM

Transmission electron microscopy (TEM) has been a paramount technique in the characterization of materials at the nanometre scale. Since the first TEM built by Max Knoll and Ernst Ruska in 1931, there have been giant leaps in their performance. With the development of Field Emission Guns in the 1970s, Scanning Transmission Electron Microscope (STEM) systems with a resolution of ∼10 nm was achieved; it had the advantage of increased current densities compared to tungsten or LaB_6 sources.[1] TEMs soon became able to achieve high resolution on a routine basis. The first spherical aberration corrector was developed in 1997 and sub Å resolution achieved in 2002.[2,3] Aberration correction opened up a whole new world of possibilities in the characterization of nanomaterials. Researchers were able to probe into the nanoworld with the use of smaller and more intense electron beams, enabling the acquisition of spectroscopic data at the atomic scale. Aberration correctors can correct either the probe-forming lens for the best performance in STEM, or the image-forming lens for the best performance in TEM mode. In STEM mode, a focused probe is rastered across an electron transparent sample

[a]*Department of Advanced Electron Microscopy, Imaging and Spectroscopy, International Iberian Nanotechnology Laboratory, Avenida Mestre Jose Veiga, Braga 4715-310, Portugal. E-mail: leonard.francis@inl.int*
[b]*UOW Electron Microscopy Centre, University of Wollongong, Wollongong, New South Wales, Australia*

generating many different signals, which form different types of images. The resolution of the image depends mainly on the probe size, which is limited by the positive spherical aberration introduced by round lenses. However, aberration correctors can introduce a negative spherical aberration coefficient reducing the probe size to sub-angstrom dimensions. Bright Field images are formed using electrons scattered at very low angles, typically from 0 to <15 mrad. Bright field images contain diffraction information and are equivalent to TEM imaging, through the principle of reciprocity.[4] The most widely used techniques, include angular dark field (ADF) imaging, which consists of collecting and imaging the forward scattered electrons (>15 mrad) and high angle annular dark field (HAADF), which collects and images the electrons scattered at much higher angles (typically >50 mrad). HAADF images are incoherent in nature, therefore, the signal intensity is proportional to thickness and the atomic number of the atoms in the specimen.[5] Since TEM specimens are usually homogenous in thickness, the contrast in HAADF images originates from a Z dependence, with high Z areas showing brighter contrast and *vice versa*. Recent works have shown that SE imaging in the TEM can be very useful to correlate topography information with transmission information, and even to resolve the atomic structure of reconstructed surfaces.[6] Chemical identification is also possible by collecting the characteristic X-rays emitted by the specimen, a technique called energy dispersive X-ray spectroscopy (EDS). However, EDS is limited to chemical composition due to the limited energy resolution, typically ~120 eV. Electron energy loss spectroscopy (EELS) is a technique that measures the energy transferred from the main e-beam to the specimen as the beam passes through the sample. Resolution in EELS ranges from ~0.7 eV for FEG instruments down to ~9 meV with the aid of monochromators.[7] With higher energy resolution it is possible to not only get chemical composition but to study bonding, electronic states, surface plasmons, and more.[8–10]

In TEM mode, the electron beam is projected nearly parallel to the specimen plane. After interacting with the sample, the Objective Lens focuses the diffracted beams to form a TEM image in the front focal plane of the first transfer lens, which then passes through the intermediate lenses magnifying the image millions of times to record it in a charge-coupled device (CCD). Similar to STEM, the limitation in TEM resolution is due to the spherical aberration of the image-forming lens, which can be compensated with an aberration corrector. The advantage of TEM over STEM is that the images are acquired in parallel, which means faster acquisition times and less sources of artefacts. Faster image acquisition makes TEM more suitable to study dynamic processes. On the downside, TEM is inherently unable to create EDS spectrum images, so it is mostly limited to imaging and Energy Filtered TEM. Both TEM and STEM provide substantial information of materials at the nanoscale, and work as complementary techniques to fully characterize phenomena at the nanoscale.

In this chapter, the recent advances in the applications of aberration corrected S/TEM (AC S/TEM) combined with spectroscopic techniques in the study of nanomaterials are highlighted. The examples include carbon-based materials, nanoparticles, 1D and 2D Materials, beam sensitive materials, *etc.* The techniques covered include 3D spectroscopic tomography, HAADF-STEM imaging, *in situ* electron microscopy, to name a few. *In situ* TEM experiments push the limits of "conventional" AC S/TEMs by studying fully dynamic processes; from catalytic reactions, optical properties, to mechanical deformation, *etc.*[11–13] TEM-based studies are becoming more and more popular every year due to the information-rich experiments that are performed, in a bid to solve important questions in many different fields, ranging from catalysis to semiconductors. This chapter intends to cover some of these latest developments as well.

2 AC-TEM imaging of carbon nanomaterials and related hybrid structures

With the advent of AC S/TEMs, it is now routinely possible to image C-based nanostructures and obtain unprecedented information not possible previously.[14–17] However, imaging of these nanostructures is preferably carried out at low voltages, typically below the threshold value for C (86 kV) in order to avoid electron irradiation-induced damages that may alter the specimen under observation. Hence, 80 kV is typically used for observation of C nanomaterials such as Graphene and Carbon Nanotubes (CNTs). In fact, 60 kV is preferred since even at 80 kV dynamics of the C nanostructures is not completely ruled out. An immediate consequence of low-voltage imaging is the effect of chromatic aberrations (C_c) that becomes dominant at lower voltages. The defocus spread due to chromatic aberration is given by $C_c = dE/E_o$, where C_c is the chromatic-aberration coefficient of the lens, dE is the energy loss of the electrons and E_o is the initial beam energy. In order to overcome C_c and achieve an information limit better than $(0.1\ \text{nm})^{-1}$ TEMs are additionally equipped with a monochromator.[18–21]

2.1 Molecules and molecular interactions in C nanomaterials (CNTs and Graphene)

Real time imaging of molecular transformations with atomic resolution was achieved by AC-TEM, revealing the chemisorption of aromatic molecules on pristine graphene.[22] The TEM imaging in this case provided insights on the reversible transformations between ring- and line-like structures of individual perchlorocoronene ($C_{24}Cl_{12}$) molecules on pristine graphene. The electron beam promoted the formation of free radicals of $C_{24}Cl_{12}$ molecules and their covalent attachment to the graphene substrate. Anchoring of monomeric and dimeric pyrenes on the sidewalls of single-walled carbon nanotubes (SWNTs) was identified by carrying out AC-TEM imaging at 60 kV. The visualization of such molecules at the atomic scale was achieved for the first time, providing insights into the structure and dynamics of single molecules and their transformations.[23] The time evolution of conformational changes of single molecules anchored onto the surface of Carbon Nanohorns (CNHs) was followed between 120 kV to 60 kV and it was found that the magnitude of the accelerating voltage altered the frequency of conformational changes, with the motion becoming more frequent when lowering the voltages.[24] AC-TEM at 80 kV was employed to image water confined between two graphene sheets. The water molecules adopt a square "ice" structure with 90 degrees hydrogen bonding within the layers, in comparison to bulk ice phases that exhibit tetrahedral coordination of hydrogen bonds (Fig. 1).[25]

Although ≤80 kV is preferred for C based and hybrid systems as highlighted with the examples above, the question arises as to whether it tends to be the suitable choice of voltage when there is a need to do simultaneous imaging of supported metal atoms or clusters on C. This was clarified recently by the report on the tandem imaging of polyoxometalate clusters on supported graphene oxide where the choice of the voltage varied between 200 kV and 80 kV in order to obtain complementary information of both the cluster and the support.[26]

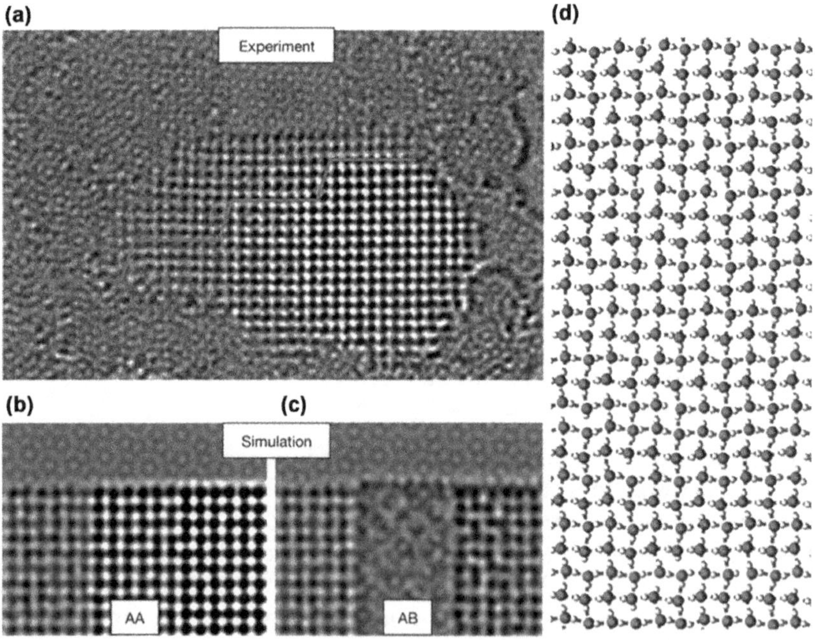

Fig. 1 Few-layer ice and its stacking order: (a) Isolated crystallite with a varying number of layers. The moiré pattern (seen clearly at the top) is due to encapsulating graphene. The curve shows changes in the contrast averaged over the corresponding parts of the image. The changes occur in quantized steps. (b) and (c) Simulated TEM images for monolayer, bilayer and trilayer ice with AA and AB stacking, respectively. The AA stacking agrees well with the experimental images whereas the AB stacked ice results in the qualitatively different appearance. (d) Typical snapshot of MD-simulated water in a graphene nanocapillary. Circles show oxygen and hydrogen; short lines indicate hydrogen bonds within H_2O. Reprinted by permission from Macmillan Publishers Ltd: Nature (G. Algara-Siller, O. Lehtinen, F. C. Wang, R. R. Nair, U. Kaiser, H. A.Wu, A. K. Geim and I. V. Grigorieva, Nature, 2015, **519**, 443), Copyright 2015.

2.2 Filled carbon nanotubes

The spatial confinement of atoms, molecules and crystal phases within nanotubes has been particularly exploited for a wide range of purposes, including synthesizing new hybrid nanostructures, modulating the photophysical and electronic characteristics of SWCNTs, and improving the stability and robustness of the chemical and photophysical properties of the guest components. The electronic properties of SWCNTs can be modified *via* the encapsulation of donor or acceptor molecules. In the case of fluorescent molecules for example, their encapsulation within SWCNTs can be expected to prevent aggregation-caused quenching.[27,28] AC-TEM imaging in combination with EELS has enabled the observation of quite interesting phenomena, including host–guest charge transfer for example in these nanotube hybrids. Theoretical studies have shown that subsequent to the encapsulation of long linear carbon chains (LLCCs) within double walled carbon nanotubes (DWNTs), the resulting hybrid systems LLCCs@DWNTs show metallic behaviour due to charge transfer from the host to the guest, despite the fact that the individual systems themselves are semiconducting in nature. Recently direct imaging of LLCCs by AC-TEM in combination with image simulations have confirmed their encapsulation within DWNTs.[29] Inorganic reactions of reactants encapsulated within SWNTs (which function as nanoreactors), have been followed by employing AC-TEM imaging. Thus the sequential encapsulation of metal carbonyls and their transformation into metal iodides, and finally into 2D transition metal chalcogenide (TMDCs) nanoribbons, has been followed throughout by carrying out imaging at each step. The SWNTs in this case act as a reservoir of electrons and aid in these reactions.[30]

AC-TEM served the dual purpose of imaging as well as to induce chemical transformations in the case of group VIII metal clusters encapsulated within SWNTs. A recent report revealed, that depending on the nature of the transition metal (Os *vs.* Ru *vs.* Fe) and its corresponding influence with the varying energies of the electron beam (80 keV *vs.* 40 keV *vs.* 20 keV), two different mechanistic pathways, namely metal-assisted e-beam induced ejection (EBIE) or e-beam induced restructuring (EBIR), can occur in the case of these SWNT@MC (MC – metal carbonyls) hybrids. The key highlights of this study, include visualizing real time alterations of nanostructures induced by the E-beam, providing insights into the nature of the metal-C bonding and their dynamics, as well as mechanisms associated with the metal assisted growth of nanostructures.[31] Another example of *in situ* electron beam irradiation and simultaneous imaging is provided by the encapsulation and growth of initially filled amorphous C into crystalline SWCNT encapsulated within boron nitride (BN) Nanotubes (NT) (Fig. 2). The BN NT in this case provides the protection sheath during the course of the irradiation process and the final formed BN-CNT hybrid heterostructures benefit from their complementary properties.[32]

It is clear from the discussions above; that lower voltages ensure reliable imaging and minimizes the effect of knock-on damage when imaging carbon based nanomaterials and hybrids. In the case of imaging

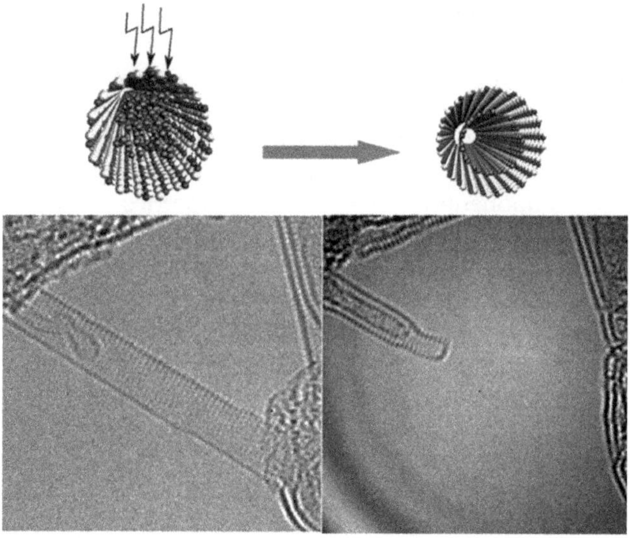

Fig. 2 Cross-sectional view of the concentric BN-C nanotubes shown along with TEM images displaying the formation process of a crystalline C-NT from disordered amorphous-like carbonaceous nanostructure encapsulated in a BN-NT under TEM radiation exposure. Reprinted with permission from R. Arenal and A. Lopez-Bezanilla, ACS Nano, 2014, 8(8) 8419-8425 and ACS Nano 2015, 9, 7731. Copyright 2015 American Chemical Society.

molecules, this is extremely critical since even if the imaging is carried out at low kV, while the behaviour is similar for all atoms, H serves to be an exception. The probability of ejection of H increases drastically at low kV. This poses a serious issue for observing molecules and molecular processes. However, this was recently circumvented by isotopic substitution whereby the substitution of H by Deuterium, enhances its stability significantly during the course of imaging. AC-TEM imaging of coronene@SWNT *vs.* coronene-d_{12}@SWNT (under identical imaging conditions) clearly showed that the former is significantly less stable under the e-beam, decomposing into amorphous featureless structures extremely fast, whereas coronene-d_{12} tends to be significantly more stable (by a factor of >2.0) during the course of imaging. Such an approach now provides an alternative route to imaging organic and biological structures.[33]

3 Nanoparticles

AC S/TEM is one of the best techniques to study inorganic nanoparticles. Their inherent nano size makes them an ideal specimen for characterization. With AC microscopes, imaging single atoms on the surfaces of the nanoparticles is readily achieved, to the point that a full 3D reconstruction of every atom is achievable.[34] However; direct visualization of nanoparticles is the most common application of the technique to enable researchers study the structure and properties of nanoparticles/clusters. Particularly, metal nanoparticles are commonly studied by AC-STEM due to the high contrast they show in HAADF imaging, in which the intensity is proportional to the atomic number by $\sim Z^{1.7}$.[5] Recent works have used this technique to study metallic NPs,[35–41] bimetallic NPs,[42–48] and metal oxides NPs.[49]

The identification of new structures at the nanoscale is one of the main objectives of characterizing NPs by AC-STEM, as shown by Mayoral *et al.*[42] In this report, HAADF imaging was employed to identify the growth of a Au plasmonic shell on a magnetic Co core of 2 nm. The Au shell was formed by 10 units that grew on the 10 facets of the icosahedron core when viewed down the 5-fold axis. Such a structure, needs chevron defects as "linking units" to fill in gaps, in what otherwise would have strained the Au lattice beyond its limits, as clearly shown by the HAADF images (Fig. 3).[42] Heavy metals are very suitable to imaging by HAADF due to the high SNR and the discrete changes in intensity between one and few atoms that enables a relatively easy 3D reconstruction.[37,39,41] To further improve the SNR, the selection of different substrates makes a big difference. If the clusters are supported on a second crystal, imaging these clusters in a hole of the carbon film support is ideal, however, the crystal support will add signal to the background. On the other hand, graphene is the best substrate as it provides minimum background signal and is stable under the electron beam, as was shown by Miramontes *et al.* wherein single atoms were easily identifiable over the graphene lattice,

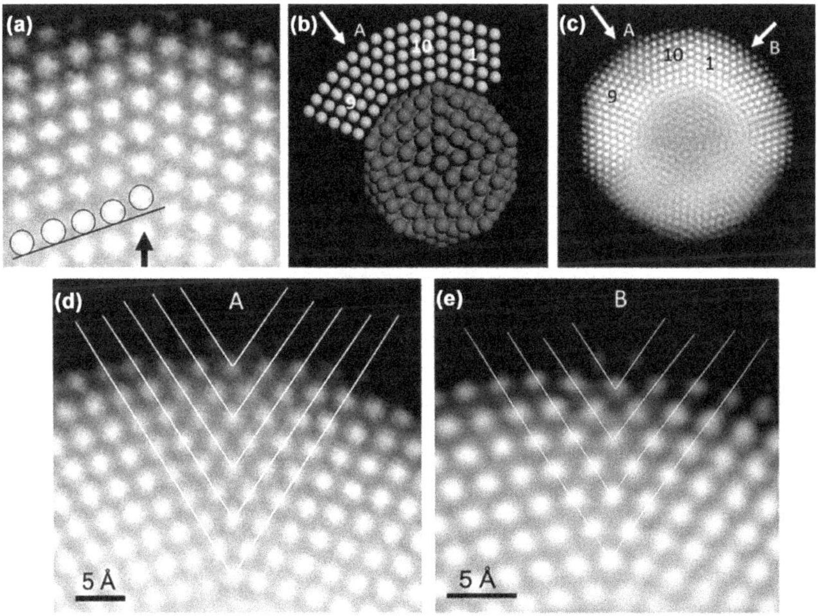

Fig. 3 C_s-corrected STEM-HAADF image of the magnified region displaying a twin plane. The atomic positions are marked by black circles; the atomic column belonging to the twin plane is denoted by a black arrow pointing towards the direction where these atomic columns are displaced. (b) Atomistic model of the Co core (dark) and Au shell (light), with one of the intergrowths marked (arrow) as A. The fcc facets are numbered. (c) C_s-corrected STEM-HAADF image of a core@shell nanoparticle where two intergrowth positions are marked as "A" and "B". (d) and (e) zoom of the triangular junction units. Reprinted from A. Mayoral, D. Llamosa and Y. Huttel, Chem. Commun., 2015, 51(40), 8442, with permission from the Royal Society of Chemistry.

enabling the identification of cluster geometries in a straightforward fashion.[41]

While most TEM data is 2D in nature, 3D reconstructions of NPs down to the atomic level expands the wealth of information extracted from AC-STEM analysis. This was achieved, using HAADF imaging with discrete tomography to visualize the under-coordinated surface atoms of complex nanoporous Au films in three dimensions.[50] The determination of the coordination of the surface atoms is critical to understand the underlying mechanisms of Au catalysis of the nanoporous Au film.[50] More recently, a full 3D reconstruction with atomic positions of a MgO nanocube was achieved by using a single AC-TEM image.[51] This is an outstanding achievement considering that it is even possible to differentiate columns terminated in O and Mg, including half occupancy sites.[51] It was also possible to determine the structure of both upper and lower surfaces, which makes it a very useful technique to characterize catalysts at the nanoscale, since the surface atoms is where the reactions occur. While this is not the first full 3D atomistic reconstruction method for nanoparticles, it remains to be seen if this is a robust and practical method that other researchers can now apply on a routine basis.

Coalescence of nanoparticles is a phenomenon that can be readily studied by TEM. AC-TEM increases the resolution of the microscopes up to the point to detect single atoms in particular situations. As mentioned before, heavy atoms on a light substrate present an ideal system for such a TEM study. Thus, it was shown that the coalescence of Au atoms on a thin C support can be studied with single atom sensitivity. The contrast in the AC-TEM images was enough to observe that single atoms in-between particles act as catalysts that accelerate the coalescence process.[52]

AC-STEM has been a great resource in characterizing supported catalysts. Lin *et al.* reported a very good case of how HAADF imaging, can be very useful in studying the adhesion and atomic structures of metal particles under 5 nm on crystalline supports; Au NPs on ceria crystals specifically and their final morphology after water–gas shift reaction (WGS).[53] By analysing HAADF images, it was possible to determine that Au grows with extended islands (according to Stranski-Krastov mode) on a {200} ceria surface and after the WGS reaction, the extended islands at the interface would migrate to the particles increasing the average particle size and changing their morphology. Ceria {111} surfaces presented a different story. Particles would grow without extended islands and individual rafts were visible on the {111} surfaces, however, after the WGS reaction the rafts migrated to the nanoparticles increasing their size without a significant change in their morphology. This shows the size, morphology and atomic interface dependence on the ceria support, namely different surfaces, and their reaction environment.[53]

Another advantage of HAADF imaging is that it does not suffer from diffraction contrast; hence, it is an ideal technique to reconstruct 3D volumes by a tomographic tilt series acquisition. While this method is time consuming, typically acquiring more than 100 images of the same particle at different tilts, the information provided can be of paramount importance. A good example is the 3D reconstruction of metal catalysts on a crystalline support where it is possible to identify the location of the NPs in the catalyst, for example at a grain boundary,[45] or in another case, to ascertain if the NPs were distributed inside a bigger structure such as a zeolite.[54]

4 1D Nanomaterials

Ultrathin Au nanowires (<2 nm) investigated by AC-TEM have shown to have a highly defective stacking sequence of atomic planes which start to evolve under the electron beam and finally undergo necking and breaking. A detailed analysis carried out by imaging, and supported by simulations and first principles calculation, reveal that the nanowires have different defects, such as intrinsic stacking faults (ISF), extrinsic stacking faults (ESF), and twin boundaries (TB). These defects behave differently and they evolve according to their surface energy. The breakdown process and the final morphology was identified to be mainly due to strain effects (Fig. 4).[55] Monochromated STEM-EELS was employed to map the plasmons of the new metastable phase of Au, namely the 4H polytype in the case of Au nanoribbons (NR). The simulated and observed EELS spectra on a single Au NR, demonstrate that the optical response of 4H Au is quite different to that of face centred cubic (fcc) Au, thus opening up new possibilities for the controlled synthesis of functional nanomaterials with novel plasmonic properties.[56] HAADF-STEM imaging of helical ultrathin Au–Cu nanowires of ~2 nm diameter showed the presence of Boerdijk–Coxeter–Bernal type structure. The Au–Cu system was formed with a self-assembled external coiled shape and with a high density of twin boundaries and stacking faults. The nanowires in this case showed the formation of a helicoidal structure which is generated by the oriented attachment of small octahedral (or truncated octahedral) seeds sharing {111} facets and in some cases with icosahedra, decahedra and single nanocrystals, which coexist into the synthesized solution that eventually coalesce into a single nanowire (Fig. 5).[57]

AC STEM was employed, to unambiguously reveal the presence of ultrathin Re nanowires confined within the hollow core of CNTs. The Re NWs in this case, had an unusual fcc structure not seen in the case of the bulk (hcp structure) and revealed a well-defined four- and nine-atom configurations due to the confinement effects of the CNTs. Density functional theory calculations confirmed, that the formation of unusual fcc-stacking Re nanowires is largely facilitated by the strong interaction between Re atoms and CNTs and the spatial restriction within the CNTs.[58] 1D atomic chains of Li metallofullerenes encapsulated within SWCNTs were identified with single-atom precision and mapped employing STEM-EELS. This is significant since identifying and mapping such low atomic number atoms, ex. Li ($Z = 3$) as single atoms employing TEM/STEM or EELS has not been possible until now. This difficulty arises especially due to their low scattering power as well as higher "knock-on" probability which was circumvented in this case, by encapsulating them in "carbon cages", thus making it easier to locate and map them in their confined nanospace.[59]

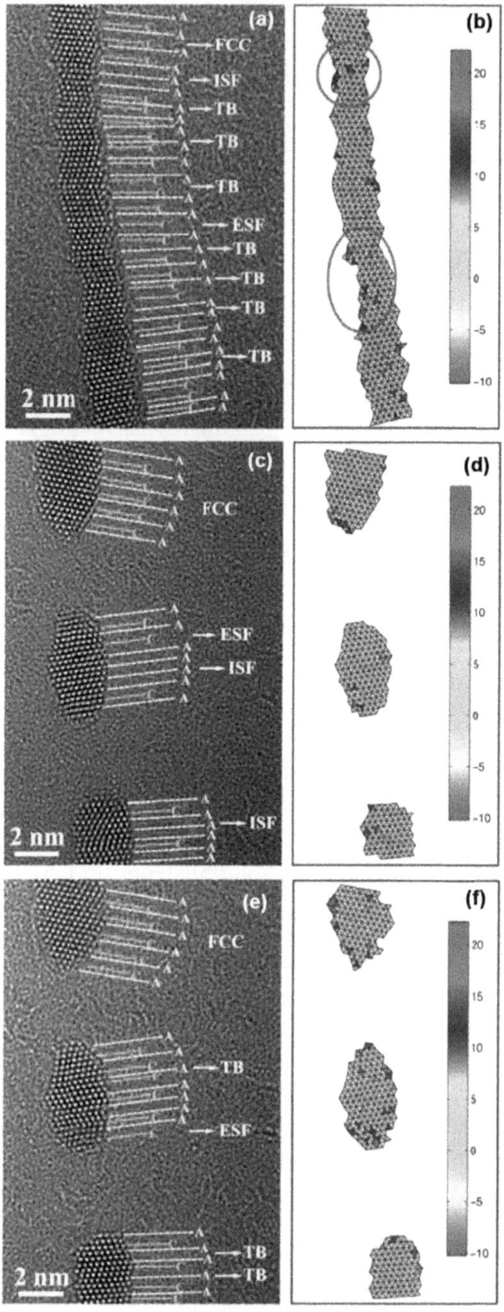

Fig. 4 Defect and strain analysis of the ultrathin Au NW. AC TEM images of the Au NW (a) before breakdown, (c) after breakdown, and (e) after further relaxation. The corresponding atomic-scale strain mappings are shown in panels b, d, and f, respectively. The scale bars indicate the stain in percentage (%). Stacking sequence of the {111} atomic planes and different types of generalized stacking faults (GSF) defects are labelled in AC-TEM panels a, c, and e. The circles in panel b indicate the necking positions with inhomogeneous strain. Reprinted with permission from Y. Yu, F. Cui, J. Sun and P. Yang, Nano Lett., 2016, 16(5), 3078. Copyright 2016 American Chemical Society.

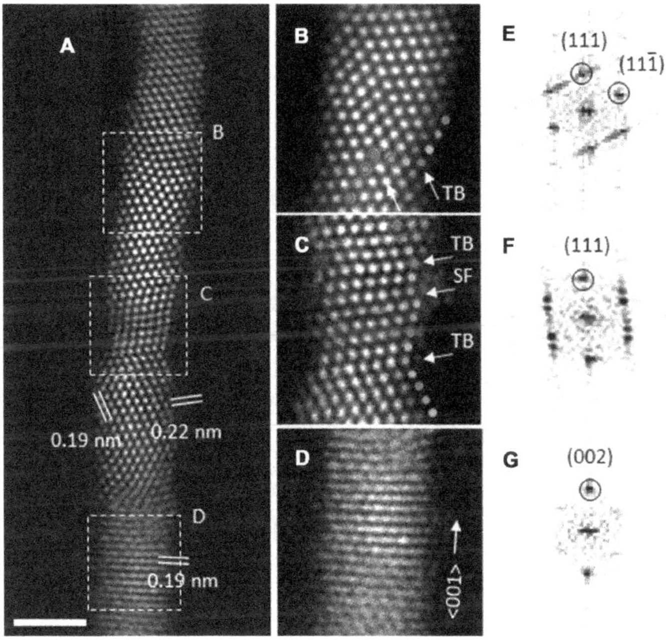

Fig. 5 (A) HAADF-STEM image of a single ultrathin helical nanowire oriented near to the $[\bar{1}10]$ axis zone. (B–D) Magnified zones with multiple defects highlighted. The atomic packing variations are marked with circles corresponding to A, B, C atomic layers, respectively. (E–G) FFT of each region is shown on the right in order to accentuate the observed defects. Scale bar: 2 nm. Reprinted with permission from R. Mendoza-Cruz, L. Bazán-Díaz, J. J. Velázquez-Salazar, G. Plascencia-Villa, D. Bahena-Uribe, J. Reyes-Gasga, D. Romeu, G. Guisbiers, R. Herrera-Becerra and M. José-Yacamán, Nano Lett., 2016, 16(3), 1568. Copyright 2016 American Chemical Society.

5 2D Materials

TEM has been a fundamental tool in the process contributing to the characterization of 2D materials. It is the only technique that allows the characterization of crystal defects,[60,61] behaviour of dopants,[8,62] among other unique properties. As stated before, particular care has to be taken to avoid introducing artefacts in the 2D structures due to radiation damage, very similar to the case of graphene. HAADF imaging has proven to be a great technique to characterize 2D materials even to the point that it can identify the chemistry of individual atoms and differentiate between low Z elements, such as B, C, N.[63] It is particularly useful for the characterization of 2D heterostructure interfaces.[64–67] Li *et al.* fabricated an atomically sharp WSe_2–MoS_2 heterostructure and used the Z sensitivity of the HAADF micrographs for easy differentiation of WSe_2 from MoS_2 due to the difference in atomic numbers, W being so heavy compared to Mo, that WSe_2 appears much brighter in the images.[64] In contrast, a more diffused interface between MoS_2 and WSe_2 was observed by Huang *et al.*; no sharp interface was readily apparent.[65] These novel heterostructures opens up new possibilities in the fabrication and applications of electronic devices, with the characterization of the interfaces being of utmost importance.

AC-STEM revealed phase transformations at the nanoscale in MoS_2 as reported by Lin *et al.*[68] In this study Re-doped MoS_2 phase transformation from semi-conductor (2H) to metallic phases (1T) using HAADF imaging at 60 kV was observed. The MoS_2 monolayers, were heated up to temperatures of 700 °C providing enough energy to the MoS_2 to undergo a phase transformation. HAADF imaging revealed that the phase transformation starts at the Re position as shown in Fig. 6. Thus, it was possible to elucidate the phase transformation mechanisms, which requires either gliding of atomic planes of S and/or Mo going through an intermediate phase, the α-phase, as precursor, or the formation and migration of two types of boundaries in the MoS_2 monolayer. This process can be visualized in the HAADF-STEM sequence in Fig. 6.[68]

The electron beam in a TEM is often used as a tool to modify/create nanostructures. Quantum wires, quantum dots and Shottky diodes were

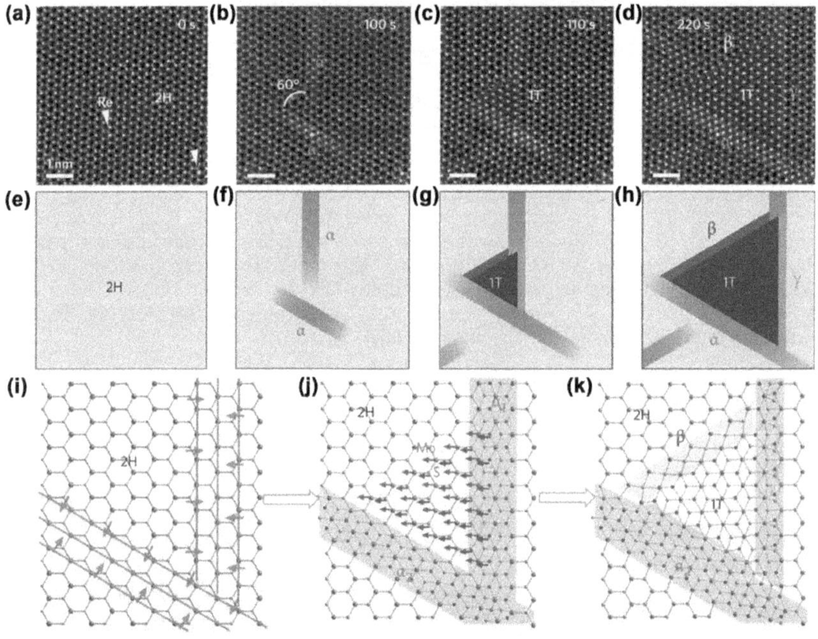

Fig. 6 Atomic movements during 2H to 1T phase transformation in single-layered MoS_2 at $T = 600$ °C. (a) Single-layered MoS_2 doped with Re substitution dopants (indicated by arrowheads) has the initial 2H phase of a hexagonal lattice structure with a clear hollow centre (HC). (b) At $t = 100$ s, two identical intermediate (precursor) phases (denoted α) form with an angle of 60°, and consist of three constricted Mo zigzag chains. (c) At $t = 110$ s, a triangular shape indicating the 1T phase (~1.08 nm^2) appears at the acute corner between the two α-phases. The 1T phase provides noticeable contrast because of the S atoms at the HC sites. (d) At $t = 220$ s, the area of the transformed 1T phase is enlarged to ~8.47 nm^2. Three different boundaries (α, β and γ) are found at the three edges between the 1T and 2H phases. (e–h) Simple schematic illustrations of the 2H to 1T phase transition corresponding to the ADF images in a–d, respectively. (i) Atomic model of α-phase formation by the constriction of three Mo zigzag chains. (j) Nucleation of the 1T phase (triangular) with the Mo + S (or S′) atoms gliding in the directions indicated by blue and pink arrows. (k) β-Boundary formation at the growth frontier side. The α_1-phase transforms to a γ-boundary, and the α_2-phase becomes wider. Reprinted by permission from Macmillan Publishers Ltd: Nature Nanotechnology (Y.-C. Lin, D. O. Dumcenco, Y.-S. Huang and K. Suenaga, Nat Nanotechnol., 2014. 9(5), 391-396), copyright 2014.

patterned onto MoS_2 nanosheets by manipulating the electron beam and tested *in situ* using the TEM as a platform to fabricate and test the nanostructures.[68] AC-STEM has also been useful in characterizing the latest synthesized 2D materials: phosphorene[69] and borophene.[70] Similar to graphene, AC-STEM has been an extremely important technique to study the structure of these novel 2D crystals.

AC-STEM is able to directly image structural defects,[61,71] as well as study their evolution[8,60,62] in 2D crystals. Lin *et al.* studied the bonding of N onto different defective sites in graphene and their absorption of metal atoms.[8] N atoms bonded with C atoms in two main configurations with different electronic structures: Graphitic-N; which consist of three sp^2 bonds C–N bonds, and pyridinic-N defects; which consist of only 2 C–N bonds in a hexagon. EELS can resolve the exact configuration of the defective regions. EELS spectra of different N atoms at different positions in the graphene revealed a weaker bond of the pyridinic-N defect. Due to this weaker bonding of the N atoms at pyridinic-N defects, they are expected to act as an absorption site for metal dopants, which was then experimentally confirmed by directly visualizing the atomic structure and confirming the elements of the dopant atoms by EELS, as shown in Fig. 7.[8] It is important to mention that the microscope used in this experiment was operated at 60 kV, to avoid the introduction or alteration of defects. Graphene has also been imaged at 30 kV by Sawada *et al.*[72] which enables minimum damage to the sample for longer periods of time.[73,74] Structural defects in 2D materials determine their properties, hence, AC-STEM is a fundamental tool to determine the types of defects present in the 2D structure.

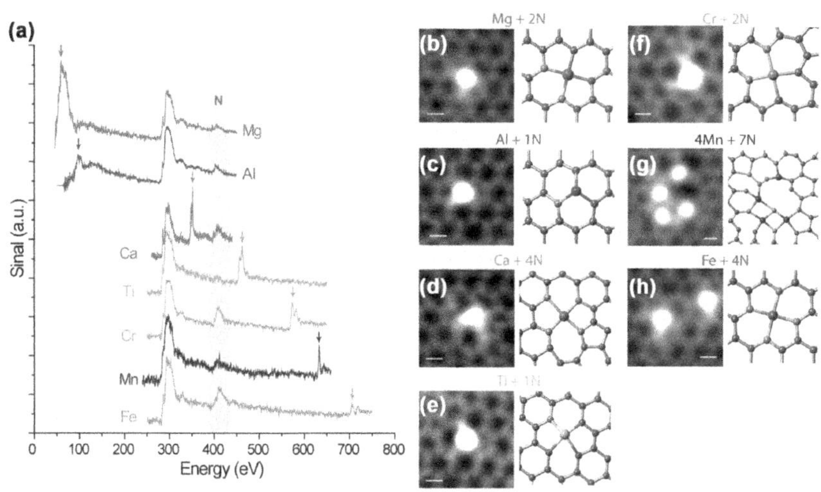

Fig. 7 (a) EEL spectra for single atoms of various metal impurities, Mg, Al, Ca, Ti, Cr, Mn, and Fe, doped in O_3+N_2 plasma treated graphene layer. All of them are found at the N-defect sites in graphene. (b)–(h) Corresponding ADF images of all the metal in N defect structures. Note that the N K-edge is always visible around 396–440 eV as well as the carbon K-edge at 285 eV. Scale bar is 2 Å. The weak N signal in the Ti + N is attributed to the area of EELS line scan, which may not completely hit the N atom. Reprinted with permission from Y.-C. Lin, P.-Y. Teng, C.-H. Yeh, M. Koshino, P.-W. Chiu and K. Suenaga, Nano Lett., 2015, 15(11), 7408. Copyright 2015 American Chemical Society.

6 Beam sensitive materials

During TEM observations there are two important mechanisms of electron beam damage of specimens, namely radiolysis and knock-on damage.[75,76] Knock-on damage is the displacement of the atoms of the specimen due to the incident electrons of the incoming beam. This process could be overcome by reducing the accelerating voltage, as for most inorganic materials there is a threshold value.[77] This was discussed previously as the main reason why C based nanostructures and related hybrids were imaged at 60 kV. Radiolysis on the other hand (ionization damage) results from the dissociation of chemical bonds due to the high energy flux onto the material. In a TEM the interactions comprise several electron excitations, such as inner shell ionization, plasmon loss, creation of locally bound electron–hole pairs and the cross section of the samples, which paradoxically decreases with increasing acceleration voltage and is strongly dependent on the radiolytic sensitivity of the material and the sample thickness. Zeolites are a set of materials that are extremely beam sensitive mainly due to radiolysis. For the particular case of AC STEM the high current focused onto a fine spot results in the formation of a hole in the zeolites; however, the main advantage of this technique relies on all electrons being concentrated into a very fine spot and only the scanned area is affected. Therefore, by having an exhaustive control of the electron beam current, new findings are possible on the characterization of molecular sieves. While low acceleration voltages (60–120 kV) helps avoid the immediate "knock-on" damage in materials, higher voltages (200–300 kV) can penetrate thicker crystals at high speeds reducing the cross section of the samples, namely reducing the interaction between the electron beam and the specimen. In addition, imaging at higher voltages has the added benefit of an improved spatial resolution. As for most highly beam sensitive materials (ex: zeolites, mesoporous materials, metal organic frameworks (MOF), *etc.*), useful data can only be extracted from the first acquired image. Using higher voltages with less electron beam–sample interaction is a valid route to the acquisition of optimal STEM data. MOF materials for instance, rapidly lose their long-range order and crystallinity. Therefore, in the imaging of these materials, 200 and 300 kV acceleration voltages are preferably used. Additionally, careful control of the electron dose, dwell time and image magnification is needed to optimize the acquisition of images without altering the sample so that the information acquired is representative of the real structure.[78]

Only recently, the first characterization of the atomic structure of a HEU-type zeolite through use of low voltage AC TEM under low dose conditions was reported. The sensitivity of this type of zeolite to the electron beam is in the typical range for natural zeolites, therefore the methodology used for TEM imaging of intact HEU-type is also applicable to many other sensitive materials. In this case in order to minimize beam damage to the sample, all images were acquired using a low electron dose in the range of 50–150 e/Å^2 s. The TEM results in the case of the Ag-HEU and Zn-HEU clearly show the presence of small Ag and Zn(O) clusters inside the zeolite structure. To further corroborate the TEM results, electron tomography was carried out, which confirmed the presence of Ag and Zn(O) atoms inside the zeolite structure, as well as on the surface. Although a certain degree of framework collapse has certainly taken place under the extended electron beam illumination needed for electron tomography, the clusters and nanoparticles seen in the reconstruction had not moved far from their initial position. With this data, it was clear that the Ag clusters were initially inside the zeolite, while the largest Ag particles were at the Ag-HEU surface.[79]

Recently, HAADF-STEM was employed to unravel the structure of sub-nanometer Ag species in fully exchanged Ag-LTA zeolites, which had been a matter of debate for over four decades. An octahedral Ag_6 structure, surrounded by a cubic arrangement containing eight Ag atoms located at framework positions, was clearly visualized within the sodalite cages of heat-treated Ag-LTA zeolites. Due to the extreme sensitivity of Ag-zeolite materials to the electron beam irradiation, relatively low electron dose was employed in this study.[80] Another recent study carried out by HAADF-STEM imaging was the case of MOFs loaded with metal NPs. MIL-101, a giant-pore MOF material, loaded with Pt nanoparticles investigated under controlled low dose conditions in combination with image simulations, revealed the position of the small clusters within the individual pores of the MOF structure. The embedding of the Pt nanoparticles is again confirmed by electron tomography, which showed a distinct ordering of the highly uniform Pt nanoparticles.[81,82] TAPO-5 and Zn-MOF-74 imaging by HAADF-STEM enabled unprecedented atomic structure revelations never seen before, such as the clear observation of the aluminophosphatefive (AFI) type framework structure. The imaging process was carried out at 300 kV, proving that under certain conditions of beam current this technique can provide invaluable information of all kinds of molecular sieves.[83]

7 3D Techniques

In the case of heterostructures such as nanoalloys, core–shell nanotubes/nanowires and/or nanocomposites, obtaining information of the 3D structure simultaneously along with chemical composition is essential to fully understand and tailor their properties for very specific applications. It is now readily possible to obtain such 3D structural and chemical information by employing spectroscopic tomography, which has emerged as an important tool of choice. This technique thus is a combination of electron tomography along with associated spectroscopy (EDS or EELS) for chemical compositional analysis, providing the complete picture in 3D, ideally suited for complex nanomaterials. This technique has been facilitated with the developments of new hardware, namely with the advent of EDX detectors with improved signal collection efficiency, as well as with new detector configurations that now employ multiple detectors located around the specimen (enabling efficient signal collection at different tilt angles). The use of the Super-X EDS (implemented by FEI in their ChemiSTEM microscopes) detector with four SDD detectors around the specimen facilitates acquisition of EDS data during the tilt series. In addition, the high collection efficiency of the detector enables the acquisition of reliable elemental maps in a few minutes and therefore the EDS tilt series can be materialized.[84]

In order to understand the structure and chemistry in 3D at the nanoscale, three dimensional chemical mapping of similar atomic number constituents, for ex: GdI_3@WS_2 nanotubes heterostructures, was achieved for the first time employing EDS Tomography.[85] HAADF-STEM tomography could reveal the three dimensional morphology of GdI_3@WS_2 nanotubes, but identification of GdI_3 filling in WS_2 is non-trivial, as the contrast in HAADF images depends on both mass and thickness. For instance, the effectively thinner region (the hollow part of the WS_2 nanotube) with a filling of a higher Z material like GdI_3 does not result in an increase in contrast with respect to the nanotube wall. A clear distinction is possible when there is a large difference in the atomic number, for instance, GdI_3 in a carbon nanotube or in a sample of uniform thickness, even when the difference in the atomic number is small. Thus in this study, EDS tomography serves as a better tool to understand the chemical composition in three dimensions as the chemical contrast is independent of such factors. The question of whether GdI_3 filling within the WS_2 nanotubes is a nanowire or a nanotube, was ascertained by employing this technique (Fig. 8).[85]

EELS Tomography has been employed to characterize the spatial distribution of Fe and Co cations in a complex $Fe_xCo_{(3-x)}O_4/Co_3O_4$ ordered mesoporous system, which is a relevant hybrid material due to the ferrimagnetic/antiferromagnetic coupling and high surface area. In this case, EELS-tomography enabled the acquisition of high resolution data

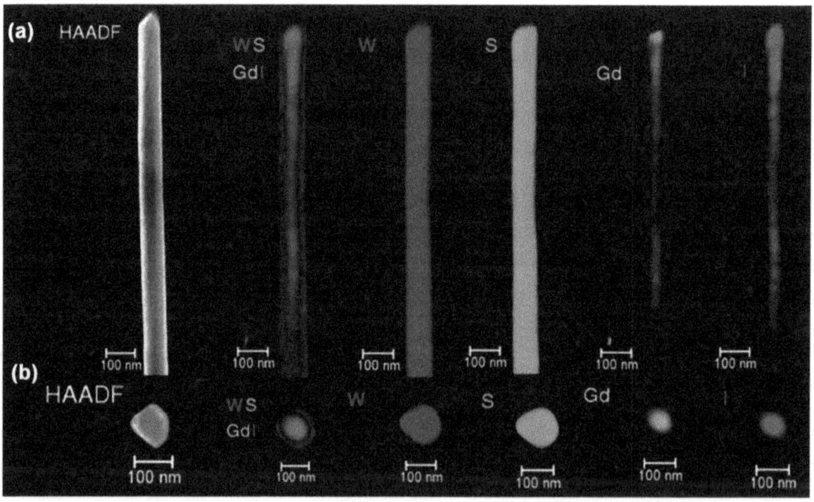

Fig. 8 Volume rendering of HAADF-STEM and EDS tomograms of GdI_3@WS_2 nanotube. (a) Nanotube positioned vertically. From left to right, HAADF-STEM tomogram, combined tomograms of each element and individual elemental tomograms are given. (b) Cross-sectional view of the nanotube. Reprinted from E. A. Anumol, A. N. Enyashin, N. M. Batra, P. M. F. J. Costa and F. L. Deepak, Nanoscale, 2016, 8, 12170, with permission from the Royal Society of Chemistry.

combined with the chemical identification, proving unambiguously that the $Fe_xCo_{(3-x)}$ does coat the inner pores of the Co_3O_4 template forming a core–shell nanocomposite.[86] Due to the energy resolution in EELS, 3D maps of the iron oxidation states of FeO/Fe_3O_4 core/shell nanocubes were successfully reconstructed from EELS-tomography.[87]

In order to improve the reliability of 3D EDX-STEM reconstructions as well as to obtain quantitative information, new 3D correction absorption effects are proposed. These improve significantly SNR and help enhance elemental mapping in 3D at the nanoscale thus enabling challenging low-dose, low-voltage experimental conditions.[88]

8 *In situ* electron microscopy

8.1 Mechanical properties of nanomaterials

In situ TEM techniques allow the mechanical properties of materials at the nanoscale to be probed while being observed in a TEM down to the atomic level with great success. Since *in situ* mechanical tests have been around for quite some time now, new phenomena at the nanometre scale continue to be discovered. Recently, it was employed to study deformation mechanisms in different systems, such as in the case of nanoscale body centered cubic (bcc) W,[89] the liquid-like behaviour of metal nanoparticles,[90–93] Au films,[94] metal NWs,[95–98] as well as the mechanical response of different semiconductor NWs.[99,100] One of the most outstanding characteristics of *in situ* TEM mechanical tests is the ability to see the deformation mechanisms down to the atomic level, revealing processes that would otherwise be impossible to unravel. The deformation mechanism of bcc W, for example, was thus recently elucidated by this technique.[89] It was found that twinning is the main deformation mechanism in bcc W under tensile loading along $\langle 100\rangle$, $\langle 110\rangle$ and $\langle 111\rangle$ directions (Fig. 9).[89] One of the main

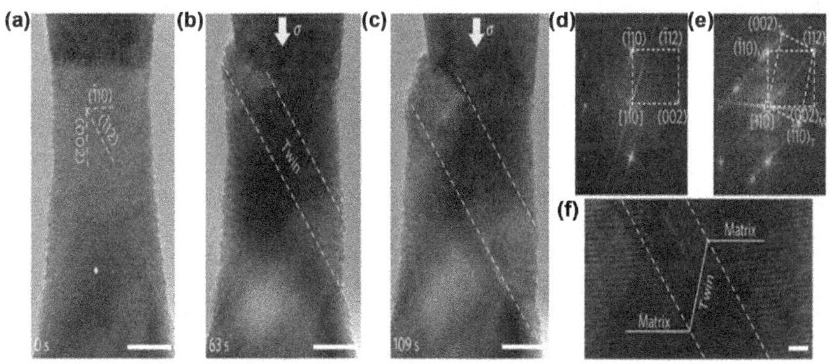

Fig. 9 Deformation twinning in a W bicrystal nanowire under [$\bar{1}$10] compression. (a–c) Sequential TEM images showing deformation twinning in a W bicrystal nanowire (15 nm in diameter) at room temperature under a strain rate of 10^{-3} s^{-1}, which is loaded along [$\bar{1}$10] and viewed along [110]. Scale bars, 5 nm. (d, e) Fast Fourier transform pattern of the pristine W bicrystal and the deformation twin, respectively. (f) Magnified TEM image showing the deformation twin. Scale bar, 1 nm. Reprinted by permission from Macmillan Publishers Ltd: Nature Materials (J. Wang, Z. Zeng, C. R. Weinberger, Z. Zhang, T. Zhu, S. X. Mao, Nat. Mater., 2015, 14(6), 594), copyright 2015.

hurdles in carrying out *in situ* work is sample preparation, which is the most difficult part. In order to overcome this difficulty, Wang *et al.* simply welded the W NWs inside the TEM by cold-welding.[89] This has the advantage that the sample NWs are in an optimal geometry, at the appropriate height, and in a good crystal orientation.

The liquid-like behaviour of metals at the nanoscale has been studied multiple times in the past by *in situ* TEM.[91,101] However, it was recently shown that the liquid-like behaviour of nm sized Ag NPs was controlled by the diffusion of surface atoms while the rest of the particle remained crystalline.[90] In this study a TEM-STM holder was used to probe the mechanical behaviour of Ag NPs under 10 nm. Using a W tip the NPs were compressed and released several times, causing significant pseudo elastic deformation; *i.e.* when the stress is applied, the atoms on the surface migrate towards the lateral surfaces to accommodate the deformation as shown clearly by the TEM images.[90] A similar phenomenon was also observed with Au NPs manipulated to form a single grain boundary.[91] The authors in this study used a Si tip to pick up a Au NP which then could be brought into contact with a second Au NP. Both particles would cold-weld and form a grain boundary.[91] It was shown that the grain boundary would change by the relative movement of the particles by surface diffusion similar to a liquid-like behaviour (Fig. 10). Moreover, this particular technique, was used to measure the surface diffusion coefficient of nm sized Au nanoparticles as was described in the study.[91]

Due to the difficulty of *in situ* TEM mechanical tests, studies usually report a single test on a particular system, or a few tests on different but related systems. However, Chen *et al.*[97] reported a systematic study of the

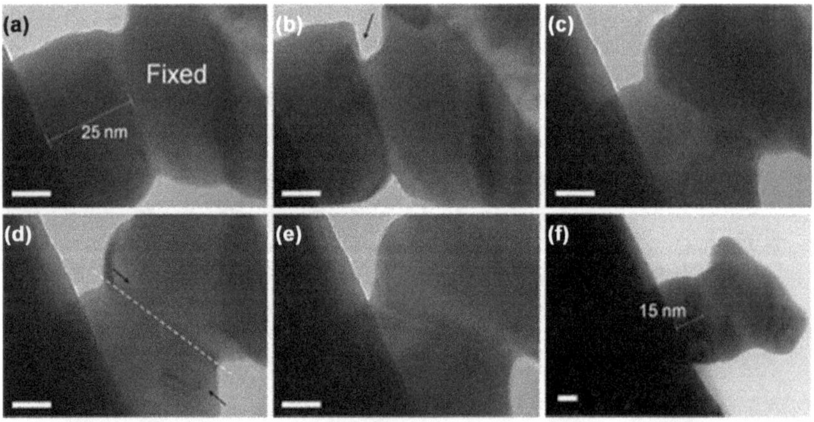

Fig. 10 Sequence of TEM images of the GB rotating and sliding. (a) GB at 69.27 s. (b) The NPs slid, leaving a surface of the bottom-left particle (marked by an arrow) (70.87 s). (c) The GB is moved back covering the free surface left open in (b) and rotates counter-clockwise (76.666 s). (d) The bottom particle is moved towards the beam direction and contrast fringes can be observed as the GB moves (marked by arrows) (80.8 s). (e) The GB is moved back to an almost on-edge view (84.47 s). (f) The NPs detached from the Si tip. (120 s). Scale bars are 10 nm. Reprinted from G. Casillas, A. Ponce, J. J. Velazquez-Salazar, M. Jose-Yacaman, Nanoscale, 2013, 5(14), 6333, with permission from the Royal Society of Chemistry.

deformation of Pd whiskers of similar size in over 60 individual whiskers. A combination of *in situ* SEM, *in situ* TEM and temperature dependant tests, were employed to determine the Young Modulus of the Pd whiskers, and its dependence to temperature, strain-rate and size. The *in situ* TEM tests revealed that dislocations mechanisms govern plastic deformation irrespective of strain-rate and temperature.[97]

Mechanical behaviour of 2D materials is an area of current interest, studied by *in situ* TEM. Although 2D materials represent a different challenge due to the high surface area, several works have been able to probe the mechanical properties of MoS_2,[102,103] and more recently, phosphorene.[104] The exfoliation of MoS_2[105] and graphene[106] has also been carried out down to single atomic layers, revealing their underlying tribological properties.

8.2 Electrical properties of nanomaterials

Measuring the electrical properties of nanomaterials is of great interest, however, to measure them in nanostructures, is not a trivial problem. Different TEM holders have been fabricated to solve this problem, either by a MEMS design or by using a W tip. A recent report by Rodríguez-Manzo *et al.* revealed the modulation of transport properties in graphene nanoribbons by *in situ* TEM electrical measurements.[107] In this study, they positioned a graphene sheet on top of a gate, source and drain electrodes, and subsequently used the electron beam in the TEM to modify the thickness of the graphene strip between the source and the drain with nm precision, demonstrating the unique possibilities of *in situ* TEM experiments. In another similar work, Wang *et al.* studied the degradation of free-standing heterostructures of graphene–WSe_2 by thermal loading and charge flow, which is suitable for any 2D structure.[108] They found that under pure thermal load, structure degradation starts at 850 °C by Se loss. On the other hand, under both electrical and thermal load the structure starts to degrade at 690 °C.

Torre *et al.* pushed the limits of the technique by performing electrical measurements on single atomic C chains, better known as carbynes.[109] In this case they were able to observe a metallic to semiconductor transition behaviour that occurs due to strain in carbynes. When the carbynes are strain-free, the structure becomes a cumulene-type, *i.e.* double bonding between the C atoms in the chain, and the metallic behaviour is observed. Contrarily, when the carbynes are strained the structure becomes plyyne-type, *i.e.* alternating single triple bonds, and the semiconductor behaviour is observed.[109] Carbyne has been proven to be stable at lengths of ~5 nm, its synthesis and imaging is difficult at best,[110] which is why being able to perform actual measurements on individual carbynes is a great achievement.

8.3 Cathodoluminesence in the TEM

In the search for higher resolution, implementing *in situ* TEM techniques has been a challenge due to the ever-closing gap in the pole-piece. In the early 1980's, Pennycook's development of the CL-STEM system was quite successful[111] up to the point to obtain CL signal from single dislocations.[112] However, it was not until the work of Zagonel *et al.*, that the latest design developments in the STEM systems were exploited and spatial atomic resolution and CL mapping was achieved.[113] In this case, the system consisted of a modified Cs corrected VG STEM with a parabolic mirror inserted between the electron source and the sample. The collection angle was ~1.4 sr, allowing enough signal to perform CL spectrum images of individual QDisks and map their optical spectra. QDisks are systems that mimic the behaviour of atoms for particular applications. Even a one unit cell difference can change the optical response of the system. This set up was enough to correlate the exact number of unit cells of the QDisks to the CL signal (though aposteriori) which was unprecedented until then. It was shown that the CL emission spectra changes dramatically from 4 to 6 and 9 mono layers. For the 4 mono layers QDisk, three different peaks were present in the CL spectrum, whereas for the 6 and 9 monolayers only one peak is observed and they presented a shift towards the infrared showing the dependence of emission energy with thickness (Fig. 11).[113]

Subsequently Tizei *et al.* went on to study individual diamond nanoparticles and identified individual colour centres on different particles in the nanometer range.[114] This study brought to light new information on single photon processes, as well as measurements of physical parameters that will help model the physics of phonons/defects in different materials. Kociak and his group continued exploring the optical properties of different materials at the nanoscale including II–VI QDs,[115] InP NWs,[116] diamond nanocrystals,[117] hexagonal boron nitride,[118] InGaN alloys[119] and metal nanoparticles.[10] In the particular case of the quantum disks, it is important to note that an important result of this work is that CL provides the same information as PL and therefore can be readily used to study the optical properties of nanostructures. In fact they compared the CL and PL signal from the exact same quantum disk (smaller than 20 nm) and the spectra of both CL and PL showed the same features.[115] Moreover, they also directly compared CL to EELS signal from a single Au nanoplate with triangular shape.[10] The main advantage of CL is that the energy resolution is 1–2 orders of magnitude better, compared to a typical energy filter.[113] However, it was found that EELS works well as a complimentary technique to CL since it reveals dark modes in the system in contrast to the CL signal.

A different group in Germany made their own *in situ* TEM CL system by modifying an FEI Tecnai 20 in the same way as the VG by inserting a parabolic mirror between the electron source and the sample. Several works have been published using data from the modified Tecnai, mainly involving GaN based semiconductors systems,[12,120–124] defects in GaAs NWs[125] and in GaN nanocolumns[126] which demonstrates the versatility of the technique. Picher *et al.* developed their own in-house solution for this, even being able to get Raman measurements from it.[127] Later on Gatan Inc. released their holder for commercial TEMs (namely JEOL and FEI) to collect CL signal *in situ* TEM, making this technique available to any group with a TEM, without the need to modify their current systems.

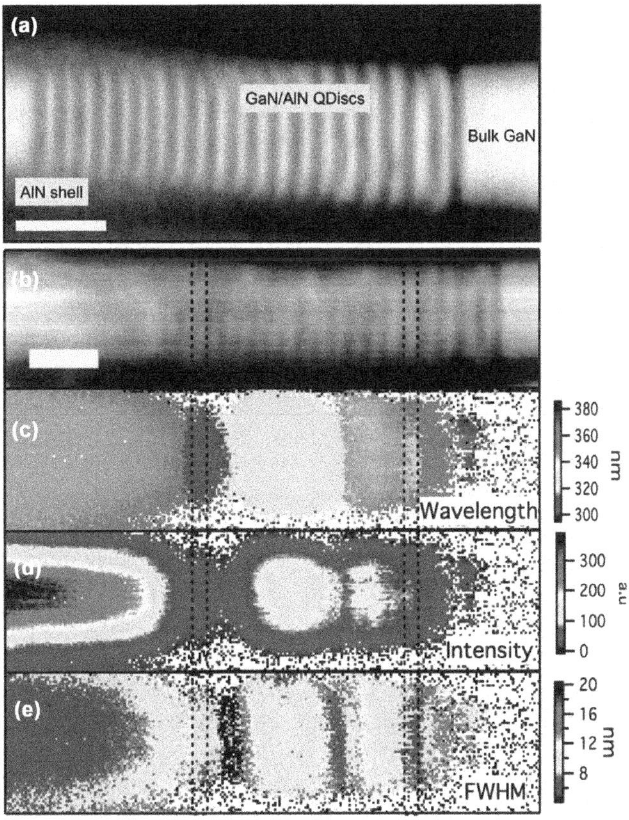

Fig. 11 Spectral imaging of a GaN/AlN NW with 20 QDisks. It was acquired with 256 per 64 pixels of spatial sampling, 0.6 nm per pixel, and with 256 pixels of spectral sampling, 2 nm per pixel (wavelength), within less than 6 min. (a) HAADF image of a GaN/AlN nanowire (GaN appears whiter, AlN darker). Scale bar is 20 nm. (b) HAADF of the NW, acquired simultaneously with the CL. Scale bar is 20 nm. (c) Wavelength position of the most intense peak. (d) Intensity of the most intense peak. (e) Full width at half-maximum of the most intense peak. Some individual spectral features can be correlated to some individual quantum disks, as emphasized by the black dashed rectangles. Reprinted with permission from L. F. Zagonel, S. Mazzucco, M. Tencé, K. March, R. Bernard, B. Laslier, G. Jacopin, M. Tchernycheva, L. Rigutti, F. H. Julien, R. Songmuang, M. Kociak, Nano Lett., 2011, 11(2), 568. Copyright 2011 American Chemical Society.

The biggest advantage of this new system is that it spans a 7.2 sr solid angle, which is at least 3 times higher than the previously in-house made solutions. A few works have proven the holder to work just as well as the in-house modified STEMs, including CL data acquisition on plasmonic nanoparticles,[128] thermoelectrics,[129] Si doped InGaN QWells,[130] $Y_2O_2S:Tb^{3+}$ and $Gd_2O_2S:Tb^{+3}$ nanometer sized phosphor crystals,[131] QDots made of a CdSe core and a ZnCdS shell,[132] among others.

9 Conclusions

It is no doubt that AC S/TEM is a great tool to study nanometer-sized structures. In 2D materials, HAADF imaging is essential in the determination of the atomic structure, defects and dopants due to the Z dependence of the signal. With a combination of HAADF imaging and EELS, it is possible to determine the bonding of atoms at different sites in a variety of crystal structures. In nanoparticles, HAADF imaging can easily identify differences in chemical composition and reveal the crystal structure, including defects at the interface and the surface. It is even possible to reconstruct the 3D structures and determine the position of every atom either by HAADF or TEM imaging. AC-STEM proves to be particularly useful to image beam sensitive materials, namely zeolites, MOFs, *etc.* In particular, HAADF imaging reveals the position of individual atoms inside these mesoporous materials by reducing the total electron-dose to the specimen, minimizing radiation damage and extracting interpretable data. It is also possible to reconstruct a 3D volume by a tomographic reconstruction of particles inside a zeolite with this method. Tomography can implemented in combination with spectroscopic data in order to create 3D volumes with chemical information of the nanostructures. AC TEM on the other hand is more suited for dynamical processes, *e.g.* coalescence of metal nanoparticles, where single atoms are identified and their role in processes elucidated. Furthermore, the implementation of *in situ* techniques (*e.g.*, deformation, heating, *etc.*) allow for probing the properties of materials at the nanoscale. Deformation mechanisms are now readily understood, due to the ability of imaging the specimen with atomic resolution, while it is being subjected to a stress. Electrical measurements are possible even down to single atomic C chains, *i.e.* carbynes. Recent developments have made possible to combine not so conventional techniques with TEMs. CL spectroscopy was achieved in QDs down to 1 nm in thickness with ~10 meV energy resolution which is an order of magnitude better than conventional EELS. This opens up a new field in nanomaterials for optical applications. The current trend in TEM is certainly moving towards *in situ* capabilities thereby enabling to tackle complex problems at the nanoscale.

Acknowledgements

Francis Leonard Deepak acknowledges the financial support provided by NORTE 2020 through the European Regional Development Fund (ERDF).

References

1 A. V. Crewe, M. Isaacson and D. Johnson, *Rev. Sci. Instrum.*, 1969, **40**(2), 241.
2 P. W. Hawkes, *Ultramicroscopy*, 2015, **156**, A1–A64.
3 P. E. Batson, N. Dellby and O. L. Krivanek, *Nature*, 2002, **418**(6898), 617.
4 J. M. Cowley, *Appl. Phys. Lett.*, 1969, **15**(2), 58.
5 S. J. Pennycook, *Ultramicroscopy*, 1989, **30**(1), 58.
6 J. Ciston, H. G. Brown, A. J. D'Alfonso, P. Koirala, C. Ophus, Y. Lin, Y. Suzuki, H. Inada, Y. Zhu, L. J. Allen and L. D. Marks, *Nat. Commun.*, 2015, **6**, 7358.
7 O. L. Krivanek, T. C. Lovejoy, N. Dellby, T. Aoki, R. W. Carpenter, P. Rez, E. Soignard, J. Zhu, P. E. Batson, M. J. Lagos, R. F. Egerton and P. A. Crozier, *Nature*, 2014, **514**(7521), 209.

8 Y.-C. Lin, P.-Y. Teng, C.-H. Yeh, M. Koshino, P.-W. Chiu and K. Suenaga, *Nano Lett.*, 2015, **15**(11), 7408.
9 *Advanced Transmission Electron Microscopy: Applications to Nanomaterials*, ed. F. L. Deepak, A. Mayoral and R. E. Arenal, Springer, 2015.
10 A. Losquin, L. F. Zagonel, B. Rodríguez-González, M. Tencé, L. Scarabelli, J. Förstner, L. M. Liz-Marzán, F. J. García de Abajo, O. Stéphan and M. Kociak, *Nano Lett.*, 2015, **15**(2), 1229.
11 P. A. Crozier and T. W. Hansen, *MRS Bull.*, 2015, **40**(01), 38.
12 G. Schmidt, M. Müller, P. Veit, F. Bertram, J. Christen, M. Glauser, J.-F. Carlin, G. Cosendey, R. Butté and N. Grandjean, *Appl. Phys. Lett.*, 2014, **105**(3), 032101.
13 T. J. A. Slater, E. A. Lewis and S. J. Haigh, *Nanoscience*, *The Royal Society of Chemistry*, 2016, vol. 3, p. 168.
14 B. Freitag, G. Knippels, S. Kujawa, P. C. Tiemeijer, M. Van der Stam, D. Hubert, C. Kisielowski, P. Denes, A. Minor and U. Dahmen, in *Instrumentation and Methods*, EMC 2008, ed. M. Luysberg, K. Tillmann and T. Weirich, Springer-Verlag, Leipzig, Germany, 2008, vol. 1.
15 T. Sasaki, H. Sawada, F. Hosokawa, Y. Kohno, T. Tomita, T. Kaneyama, Y. Kondo, K. Kimoto, Y. Sato and K. Suenaga, *J. Electron Microsc.*, 2010, **59**(Supplement), S7.
16 O. L. Krivanek, N. Dellby, M. F. Murfitt, M. F. Chisholm, T. J. Pennycook, K. Suenaga and V. Nicolosi, *Ultramicroscopy*, 2010, **110**, 935.
17 F. Börrnert, A. Bachmatiuk, S. Gorantla, D. Wolf, A. Lubk, B. Büchner and M. H. Rümmeli, *J. Microsc.*, 2013, **249**, 87.
18 S. Morishita, M. Mukai, K. Suenaga and H. Sawada, *Appl. Phys. Lett.*, 2016, **108**, 013107.
19 S. Morishita, M. Mukai, T. Sasaki, K. Suenaga and H. Sawada, *J. Phys.: Conf. Ser.*, 2015, **644**, 012033.
20 R. Leary and R. Brydson, *Adv. Imag. Elect. Opt.*, 2011, **165**, 73.
21 X. Ke, C. Bittencourt and G. Van Tendeloo, *Beilstein J. Nanotechnol.*, 2015, **6**, 1541.
22 A. Markevich, S. Kurasch, O. Lehtinen, O. Reimer, X. Feng, K. Müllen, A. Turchanin, A. N. Khlobystov, U. Kaiser and E. Besley, *Nanoscale*, 2016, **8**, 2711.
23 T. Umeyama, J. Baek, Y. Sato, K. Suenaga, F. Abou-Chahine, N. V. Tkachenko, H. Lemmetyinen and H. Imahori, *Nat. Commun.*, 2015, **6**, 7732.
24 R. M. Gorgoll, E. Yücelen, A. Kumamoto, N. Shibata, K. Harano and E. Nakamura, *J. Am. Chem. Soc.*, 2015, **137**, 3474.
25 G. Algara-Siller, O. Lehtinen, F. C. Wang, R. R. Nair, U. Kaiser, H. A. Wu, A. K. Geim and I. V. Grigorieva, *Nature*, 2015, **519**, 443.
26 C. Bosch-Navarro, L. M. Perkins, R. J. Kashtiban, J. P. Rourke, I. J. Shannon and J. Sloan, *ACS Nano*, 2016, **10**(1), 796.
27 M. Tange, T. Okazaki, Z. Liu, K. Suenaga and S. Iijima, *Nanoscale*, 2016, **8**, 7834.
28 M. V. Kharlamova, M. Sauer, T. Saito, Y. Sato, K. Suenaga, T. Pichlera and H. Shiozawa, *Nanoscale*, 2015, **7**, 1383.
29 L. Shi, P. Rohringer, K. Suenaga, Y. Niimi, J. Kotakoski, J. C. Meyer, H. Peterlik, M. Wanko, S. Cahangirov, A. Rubio, Z. J. Lapin, L. Novotny, P. Ayala and T. Pichler, *Nat. Mater.*, 2016, **15**, 634.
30 A. Botos, J. Biskupek, T. W. Chamberlain, G. A. Rance, C. Stoppiello, J. Sloan, Z. Liu, K. Suenaga, U. Kaiser and A. N. Khlobystov, *J. Am. Chem. Soc.*, 2016, **138**, 8175.

31 T. Zoberbier, T. W. Chamberlain, J. Biskupek, M. Suyetin, A. G. Majouga, E. Besley, U. Kaiser and A. N. Khlobystov, *Small*, 2016, **12**, 1649.
32 R. Arenal and A. Lopez-Bezanilla, *ACS Nano*, 2014, **8**(8), 8419–8425, and ACS Nano 2015, **9**, 7731.
33 T. W. Chamberlain, J. Biskupek, S. T. Skowron, P. A. Bayliss, E. Bichoutskaia, U. Kaiser and A. N. Khlobystov, *Small*, 2015, **11**(5), 622.
34 S. Van Aert, K. J. Batenburg, M. D. Rossell, R. Erni and G. Van Tendeloo, *Nature*, 2011, **470**(7334), 374.
35 D. Yang, P. Xu, N. D. Browning and B. C. Gates, *J. Phys. Chem. Lett.*, 2016, **7**(13), 2537–2543.
36 G. Plascencia-Villa, B. Demeler, R. L. Whetten, W. P. Griffith, M. Alvarez, D. M. Black and M. José-Yacamán, *J. Phys. Chem. C*, 2016, **120**(16), 8950.
37 H. S. Al Qahtani, K. Kimoto, T. Bennett, J. F. Alvino, G. G. Andersson, G. F. Metha, V. B. Golovko, T. Sasaki and T. Nakayama, *J. Chem. Phys.*, 2016, **144**(11), 114703.
38 J. Ohyama, A. Esaki, T. Koketsu, Y. Yamamoto, S. Arai and A. Satsuma, *J. Catal.*, 2016, **335**, 24.
39 J. Li, D. Yin, C. Chen, Q. Li, L. Lin, R. Sun, S. Huang and Z. Wang, *J. Appl. Phys.*, 2015, **117**(8), 085303.
40 T. Imaoka, T. Imaoka, H. Kitazawa, W.-J. Chun and K. Yamamoto, *Angew. Chem., Int. Ed.*, 2015, **54**(34), 9810.
41 O. Miramontes, F. Bonafe, U. Santiago, E. Larios-Rodriguez, J. J. Velázquez-Salazar, M. M. Mariscal and M. José Yacaman, *Phys. Chem. Chem. Phys.*, 2015, **17**(12), 7898.
42 A. Mayoral, D. Llamosa and Y. Huttel, *Chem. Commun.*, 2015, **51**(40), 8442.
43 L. Bazan-Diaz, R. Mendoza-Cruz, J. J. Velázquez-Salazar, G. Plascencia-Villa, D. Romeu, J. Reyes-Gasga, R. Herrera-Becerra, M. José-Yacamán and G. Guisbiers, *Nanoscale*, 2015, **7**(48), 20734.
44 A. Spitale, M. A. Perez, S. Mejía-Rosales, M. J. Yacamán and M. M. Mariscal, *Phys. Chem. Chem. Phys.*, 2015, **17**(42), 28060.
45 C. W. Han, P. Majumdar, E. E. Marinero, A. Aguilar-Tapia, R. Zanella, J. Greeley and V. Ortalan, *Nano Lett.*, 2015, **15**(12), 8141.
46 V. Beermann, M. Gocyla, E. Willinger, S. Rudi, M. Heggen, R. E. Dunin-Borkowski, M.-G. Willinger and P. Strasser, *Nano Lett.*, 2016, **16**(3), 1719.
47 L. Uson, V. Sebastian, A. Mayoral, J. L. Hueso, A. Eguizabal, M. Arruebo and J. Santamaria, *Nanoscale*, 2015, **7**(22), 10152.
48 A. Londono-Calderon, A. Bruma, D. B. Uribe, A. Ponce and M. José-Yacamán, *J. Phys. Chem. C.*, 2015, **119**(43), 24621.
49 H. Du, C.-L. Jia and J. Mayer, *Chem. Mater.*, 2016, **28**(2), 650.
50 P. Liu, P. Guan, A. Hirata, L. Zhang, L. Chen, Y. Wen, Yi Ding, T. Fujita, J. Erlebacher and M. Chen, *Adv. Mater.*, 2016, **28**(9), 1753.
51 C. L. Jia, S. B. Mi, J. Barthel, D. W. Wang, R. E. Dunin-Borkowski, K. W. Urban and A. Thust, *Nat. Mater.*, 2014, **13**(11), 1044.
52 N. Wan, L. Shuang-ying, X. Jun and M. Martini, *Nanotechnology*, 2016, **27**(20), 205605.
53 Y. Lin, Z. Wu, J. Wen, K. Ding, X. Yang, K. R. Poeppelmeier and L. D. Marks, *Nano Lett.*, 2015, **15**(8), 5375.
54 J. Zecevic, G. Vanbutsele, J. P. Jong and J. A. Martens, *Nature*, 2015, **528**(7581), 245.
55 Y. Yu, F. Cui, J. Sun and P. Yang, *Nano Lett.*, 2016, **16**(5), 3078.
56 Z. Fan, M. Bosman, X. Huang, D. Huang, Y. Yu, K. P. Ong, Y. A. Akimov, L. Wu, B. Li, J. Wu, Y. Huang, Q. Liu, C. E. Png, C. L. Gan, P. Yang and H. Zhang, *Nat. Commun.*, 2015, **6**, 7684.

57 R. Mendoza-Cruz, L. Bazán-Díaz, J. J. Velázquez-Salazar, G. Plascencia-Villa, D. Bahena-Uribe, J. Reyes-Gasga, D. Romeu, G. Guisbiers, R. Herrera-Becerra and M. José-Yacamán, *Nano Lett.*, 2016, **16**(3), 1568.
58 F. Zhang, P. Ren, X. Pan, J. Liu, M. Li and X. Bao, *Chem. Mater.*, 2015, **27**(5), 1569.
59 R. Senga and K. Suenaga, *Nat. Commun.*, 2015, **6**, 7943.
60 A. W. Robertson, G.-D. Lee, K. He, Y. Fan, C. S. Allen, S. Lee, H. Kim, E. Yoon, H. Zheng, A. I. Kirkland and J. H. Warner, *Nano Lett.*, 2015, **15**(9), 5950.
61 O. Lehtinen, N. Vats, G. Algara-Siller, P. Knyrim and U. Kaiser, *Nano Lett.*, 2015, **15**(1), 235.
62 A. W. Robertson, B. Montanari, K. He, J. Kim, C. S. Allen, Y. A. Wu, J. Olivier, J. Neethling, N. Harrison, A. I. Kirkland and J. H. Warner, *Nano Lett.*, 2013, **13**(4), 1468.
63 O. L. Krivanek, M. F. Chisholm, V. Nicolosi, T. J. Pennycook, G. J. Corbin, N. Dellby, M. F. Murfitt, C. S. Own, Z. S. Szilagyi, M. P. Oxley, S. T. Pantelides and S. J. Pennycook, *Nature*, 2010, **464**(7288), 571.
64 M.-Y. Li, Y. Shi, C.-C. Cheng, L.-S. Lu, Y.-C. Lin, H.-L. Tang, M.-L. Tsai, C.-W. Chu, K.-H. Wei, J.-H. He, W.-H. Chang, K. Suenaga and L.-J. Li, *Science*, 2015, **349**(6247), 524.
65 C. Huang, S. Wu, A. M. Sanchez, J. J. P. Peters, R. Beanland, J. S. Ross, P. Rivera, W. Yao, D. H. Cobden and X. Xu, *Nat. Mater.*, 2014, **13**(12), 1096.
66 Z. Liu, L. H. G. Tizei, Y. Sato, Y.-C. Lin, C.-H. Yeh, P.-W. Chiu, M. Terauchi, S. Iijima and K. Suenaga, *Small*, 2016, **12**(2), 252.
67 H. Heo, J. H. Sung, G. Jin, J.-H. Ahn, K. Kim, M.-J. Lee, S. Cha, H. Choi and M.-H. Jo, *Adv. Mater.*, 2015, **27**(25), 3803.
68 Y.-C. Lin, D. O. Dumcenco, Y.-S. Huang and K. Suenaga, *Nat. Nanotechnol.*, 2014, **9**(5), 391–396.
69 A. Nie, Y. Cheng, S. Ning, T. Foroozan, P. Yasaei, W. Li, B. Song, Y. Yuan, L. Chen, A. Salehi-Khojin, F. Mashayek and R. Shahbazian-Yassar, *Nano Lett.*, 2016, **16**(4), 2240.
70 A. J. Mannix, X.-F. Zhou, B. Kiraly, J. D. Wood, D. Alducin, B. D. Myers, X. Liu, B. L. Fisher, U. Santiago, J. R. Guest, M. J. Yacaman, A. Ponce, A. R. Oganov, M. C. Hersam and N. P. Guisinger, *Science*, 2015, **350**(6267), 1513.
71 O. Lehtinen, H.-P. Komsa, A. Pulkin, M. B. Whitwick, M.-W. Chen, T. Lehnert, M. J. Mohn, O. V. Yazyev, A. Kis, U. Kaiser and A. V. Krasheninnikov, *ACS Nano*, 2015, **9**(3), 3274.
72 H. Sawada, T. Sasaki, F. Hosokawa and K. Suenaga, *Phys. Rev. Lett.*, 2015, **114**(16), 166102.
73 G. R. Bhimanapati, Z. Lin, V. Meunier, Y. Jung, J. Cha, S. Das, D. Xiao, Y. Son, M. S. Strano, V. R. Cooper, L. Liang, S. G. Louie, E. Ringe, W. Zhou, S. S. Kim, R. R. Naik, B. G. Sumpter, H. Terrones, F. Xia, Y. Wang, J. Zhu, D. Akinwande, N. Alem, J. A. Schuller, R. E. Schaak, M. Terrones and J. A. Robinson, *ACS Nano*, 2015, **9**(12), 11509.
74 A. Bachmatiuk, J. Zhao, S. M. Gorantla, I. G. G. Martinez, J. Wiedermann, C. Lee, J. Eckert and M. H. Rummeli, *Small*, 2015, **11**(5), 51.
75 R. F. Egerton, *Ultramicroscopy*, 2013, **127**, 100.
76 R. F. Egerton, P. Li and M. Malac, *Micron*, 2004, **35**, 399.
77 R. F. Egerton, R. McLeod, F. Wang and M. Malac, *Ultramicroscopy*, 2010, **110**, 99.
78 A. Mayoral, P. A. Anderson and I. Diaz, *Micron*, 2015, **68**, 146.
79 M. Filippousi, S. Turner, M. Katsikini, F. Pinakidou, D. Zamboulis, E. Pavlidou and G. Van Tendeloo, *Microporous Mesoporous Mater.*, 2015, **210**, 185.

80 T. Altantzis, E. Coutino-Gonzalez, W. Baekelant, G. T. Martinez, A. M. Abakumov, G. Van Tendeloo, M. B. J. Roeffaers, S. Bals and J. Hofkens, *ACS Nano*, 2016, **10**(8), 7604.
81 M. Meledina, S. Turner, M. Filippousi, K. Leus, I. Lobato, R. K. Ramachandran, J. Dendooven, C. Detavernier, P. Van Der Voort and G. Van Tendeloo, *Part. Part. Syst. Charact.*, 2016, **33**, 382.
82 K. Leus, J. Dendooven, N. Tahir, R. K. Ramachandran, M. Meledina, S. Turner, G. Van Tendeloo, J. L. Goeman, J. Van der Eycken, C. Detavernier and P. Van Der Voort, *Nanomaterials*, 2016, **6**, 45.
83 A. Mayoral, M. Sanchez-Sanchez, A. Alfayate, J. Perez-Pariente and I. Diaz, *Chem. Cat. Chem.*, 2015, **7**, 3719.
84 P. Ercius, O. Alaidi, M. J. Rames and G. Ren, *Adv. Mater.*, 2015, **27**, 5638.
85 E. A. Anumol, A. N. Enyashin, N. M. Batra, P. M. F. J. Costa and F. L. Deepak, *Nanoscale*, 2016, **8**, 12170.
86 L. Yedra, A. Eljarrat, R. Arenal, L. López-Conesa and E. Pellicer, *Analyst*, 2016, **141**, 4968.
87 P. Torruella, R. Arenal, F. de la Peña, Z. Saghi, L. Yedra, A. Eljarrat, L. López-Conesa, M. Estrader, A. López-Ortega, G. Salazar-Alvarez, J. Nogués, C. Ducati, P. A. Midgley, F. Peiro and S. Estrade, *NanoLett.*, 2016, **16**, 5068.
88 P. Burdet, Z. Saghi, A. N. Filippin, A. Borrás and P. A. Midgley, A novel 3D absorption correction method for quantitative EDX-STEM tomography, *Ultramicroscopy*, 2016, **160**, 118–129.
89 J. Wang, Z. Zeng, C. R. Weinberger, Z. Zhang, T. Zhu and S. X. Mao, *Nat. Mater.*, 2015, **14**(6), 594.
90 J. Sun, L. He, Y.-C. Lo, T. Xu, H. Bi, L. Sun, Z. Zhang, S. X. Mao and J. Li, *Nat. Mater.*, 2014, **13**(11), 1007.
91 G. Casillas, A. Ponce, J. J. Velazquez-Salazar and M. Jose-Yacaman, *Nanoscale*, 2013, **5**(14), 6333.
92 D. Alducin, G. Casillas, F. Mendoza-Santoyo, A. Ponce and M. José-Yacamán, *J. Nanopart. Res.*, 2015, **17**(5), 1.
93 G. Casillas, J. P. Palomares-Báez, J. L. Rodríguez-López, J. Luo, A. Ponce, R. Esparza, J. J. Velázquez-Salazar, A. Hurtado-Macias, J. González-Hernández and M. José-Yacaman, *Philos. Mag.*, 2012, **92**(35), 4437.
94 E. Hosseinian, M. Legros and O. N. Pierron, *Nanoscale*, 2016, **8**(17), 9234.
95 S. Narayanan, G. Cheng, Z. Zeng, Y. Zhu and T. Zhu, *Nano Lett.*, 2015, **15**(6), 4037.
96 Y. Yue, Q. Zhang, Z. Yang, Q. Gong and L. Guo, *Small*, 2016, **12**(26), 3503.
97 L. Y. Chen, M.-r. He, J. Shin, G. Richter and D. S. Gianola, *Nat. Mater.*, 2015, **14**(7), 707.
98 D. Alducin, R. Borja, E. Ortega, J. J. Velazquez-Salazar, M. Covarrubias, F. M. Santoyo, L. Bazán-Díaz, J. E. Sanchez, N. Torres, A. Ponce and M. José-Yacamán, *Scr. Mater.*, 2016, **113**, 63.
99 Y. Chen, T. Burgess, X. An, Y.-W. Mai, H. H. Tan, J. Zou, S. P. Ringer, C. Jagadish and X. Liao, *Nano Lett.*, 2016, **16**(3), 1911.
100 K. Zheng, Z. Zhang, Y. Hu, P. Chen, W. Lu, J. Drennan, X. Han and J. Zou, *Nano Lett.*, 2016, **16**(3), 1787.
101 A. P. Merkle and L. D. Marks, *Wear*, 2008, **265**(11–12), 1864.
102 G. Casillas, U. Santiago, H. Barrón, D. Alducin, A. Ponce and M. José-Yacamán, *J. Phys. Chem. C*, 2015, **119**(1), 710.
103 D.-M. Tang, D. G. Kvashnin, S. Najmaei, Y. Bando, K. Kimoto, P. Koskinen, P. M. Ajayan, B. I. Yakobson, P. B. Sorokin, J. Lou and D. Golberg, *Nat. Commun.*, 2014, 5.

104 F. Xu, H. Ma, S. Lei, J. Sun, J. Chen, B. Ge, Y. Zhu and L. Sun, *Nanoscale*, 2016, **8**(28), 13603.
105 J. P. Oviedo, S. Kc, N. Lu, J. Wang, K. Cho, R. M. Wallace and M. J. Kim, *ACS Nano*, 2015, **9**(2), 1543.
106 G. Casillas, Y. Liao, M. Jose-Yacaman and L. D. Marks, *Tribol. Lett.*, 2015, **59**(3), 1.
107 J. A. Rodríguez-Manzo, Z. J. Qi, A. Crook, J.-H. Ahn, A. T. C. Johnson and M. Drndić, *ACS Nano*, 2016, **10**(4), 4004.
108 B. Wang, S. M. Eichfield, D. Wang, J. A. Robinson and M. A. Haque, *Nanoscale*, 2015, 7(34), 14489.
109 A. La Torre, A. Botello-Mendez, W. Baaziz, J. C. Charlier and F. Banhart, *Nat. Commun.*, 2015, 6.
110 G. Casillas, A. Mayoral, M. Liu, A. Ponce, V. I. Artyukhov, B. I. Yakobson and M. Jose-Yacaman, *Carbon*, 2014, **66**, 436.
111 H. P. Strunk, M. Albrecht and H. Scheel, *J. Microsc.*, 2006, **224**(1), 79.
112 S. J. Pennycook, L. M. Brown and A. J. Craven, *Philos. Mag. A*, 1980, **41**(4), 589.
113 L. F. Zagonel, S. Mazzucco, M. Tencé, K. March, R. Bernard, B. Laslier, G. Jacopin, M. Tchernycheva, L. Rigutti, F. H. Julien, R. Songmuang and M. Kociak, *Nano Lett.*, 2011, **11**(2), 568.
114 L. H. G. Tizei and M. Kociak, *Phys. Rev. Lett.*, 2013, **110**(15), 153604.
115 Z. Mahfoud, A. T. Dijksman, C. Javaux, P. Bassoul, A.-L. Baudrion, J. Plain, B. Dubertret and M. Kociak, *J. Phys. Chem. Lett.*, 2013, **4**(23), 4090.
116 L. H. G. Tizei, L. F. Zagonel, M. Tencé, O. Stéphan, M. Kociak, T. Chiaramonte, D. Ugarte and M. A. Cotta, *J. Phys: Condens. Matter.*, 2013, **25**(50), 505303.
117 S. Meuret, L. H. G. Tizei, T. Cazimajou, R. Bourrellier, H. C. Chang, F. Treussart and M. Kociak, *Phys. Rev. Lett.*, 2015, **114**(19), 197401.
118 R. Bourrellier, M. Amato, L. H. Galvão Tizei, C. Giorgetti, A. Gloter, M. I. Heggie, K. March, O. Stéphan, L. Reining, M. Kociak and A. Zobelli, *ACS Photonics*, 2014, **1**(9), 857.
119 G. Tourbot, C. Bougerol, F. Glas, L. F. Zagonel, Z. Mahfoud, S. Meuret, P. Gilet, M. Kociak, B. Gayral and B. Daudin, *Nanotechnology*, 2012, **23**(13), 135703.
120 M. Matin Sadat, M. Marcus, H. Jana, Z. Hao, H. W. Hergo, V. Peter, B. Frank, C. Jürgen and W. Andreas, *Jpn. J. Appl. Phys.*, 2016, **55**(5S), 05FJ09.
121 G. Schmidt, C. Berger, P. Veit, S. Metzner, F. Bertram, J. Bläsing, A. Dadgar, A. Strittmatter, J. Christen, G. Callsen, S. Kalinowski and A. Hoffmann, *Appl. Phys. Lett.*, 2015, **106**(25), 252101.
122 Ž. Gačević, N. Vukmirović, N. García-Lepetit, A. Torres-Pardo, M. Müller, S. Metzner, S. Albert, A. Bengoechea-Encabo, F. Bertram, P. Veit, J. Christen, J. M. González-Calbet and E. Calleja, *Phys. Rev. B*, 2016, **93**(12), 125436.
123 G. Schmidt, P. Veit, M. Wieneke, F. Bertram, A. Dadgar, A. Krost and J. Christen, *Phys. Status. Solidi B*, 2016, **253**(1), 73.
124 G. Schmidt, M. Müller, P. Veit, F. Bertram, J. Christen, M. Glauser, J.-F. Carlin, G. Cosendey, R. Butté and N. Grandjean, *Appl. Phys. Lett.*, 2014, **105**(3), 032101.
125 L. Bernhard, M. Marcus, W. Julia, V. Peter, R. Daniel, A. Gerhard, J. F. Jonathan, B. Frank, C. Jürgen and K. Gregor, *New J. Phys.*, 2016, **18**(6), 063009.
126 A. Urban, M. Müller, C. Karbaum, G. Schmidt, P. Veit, J. Malindretos, F. Bertram, J. Christen and A. Rizzi, *Nano Lett.*, 2015, **15**(8), 5105.

127 M. Picher, S. Mazzucco, S. Blankenship and R. Sharma, *Ultramicroscopy*, 2015, **150**, 10.
128 E. Ringe, C. J. DeSantis, S. M. Collins, M. Duchamp, R. E. Dunin-Borkowski, S. E. Skrabalak and P. A. Midgley, *Sci. Rep.*, 2015, **5**, 17431.
129 M. Zhao, M. Bosman, M. Danesh, M. Zeng, P. Song, Y. Darma, A. Rusydi, H. Lin, C.-W. Qiu and K. P. Loh, *Nano Lett.*, 2015, **15**(12), 8331.
130 J. T. Griffiths, S. Zhang, B. Rouet-Leduc, W. Y. Fu, A. Bao, D. Zhu, D. J. Wallis, A. Howkins, I. Boyd, D. Stowe, M. J. Kappers, C. J. Humphreys and R. A. Oliver, *Nano Lett.*, 2015, **15**(11), 7639.
131 J. Silver, X. Yan, G. R. Fern and N. Wilkinson, *J. Phys: Conf. Ser.*, 2015, **619**(1), 012049.
132 G. R. Fern, J. Silver and S. Coe-Sullivan, *J. Soc. Inf. Disp.*, 2015, **23**(2), 50.

Gold fabricated core–shell nanoparticles as innovative cancer therapeutic strategies to improve drug delivery

R. Dunpall*[a] and N. Revaprasadu[b]
DOI: 10.1039/9781782620358-00057

Nanobiotechnology has generated huge interest in cancer therapeutics and diagnostic medicine, as it has the potential to resolve the harmful side effects associated with conventional chemotherapies. Owing to the remarkable fundamental properties of biocompatible gold nanoparticles, the assembly and application of gold containing core–shell nanoparticles has shown promise in advancing the drug delivery applications in cancer therapy. The assembly of these unique smart architectured heterostructures developed from the need to create new and diverse materials that offer multiple functionalities. The application of gold containing core–shell nanoparticles has evolved into opportunities that far exceed the boundaries of laboratory animal experiments. Currently there are several FDA approved gold based nanoparticle drug delivery systems that are already utilised to improve healthcare and disease management protocols. This chapter therefore highlights the recent advances in the synthesis of various gold containing core–shell nanoparticles, drug loading and entrapment efficiencies, drug conjugation, functionalisation and surface responsive drug targeting.

1 Introduction to cancer therapeutics

Cancer can be defined as a collection of abnormal, damaged or mutated cells that are able to rapidly proliferate and evade the normal homeostatic and regulatory capacity of cell biology.[1–4] In order to systematically develop innovative cancer therapeutic strategies it is essential to understand the process of how a normal cell develops into a cancer cell, therefore knowledge on the biochemical regulation of cell growth cycle must be established. The human body is composed of trillions of cellular units. These cells are highly specialised units that associate to form functional tissues and organs.[3] Normal physiological processes enable the cells to grow and divide as it is required. As the cells age or become damaged they are eliminated by programmed cell death, a process known as apoptosis and that cell is replaced through a carefully regulated mitotic process that generates new cellular components and daughter cells.[1,5] Each year epidemiologists disclose alarming statistics that indicate both the incidence and prevalence of the global cancer burden is increasing profoundly. According to the International Cancer Control it is projected that by the year 2025 the global cancer burden would reach 19.3 million new cases reported each year.[6] It is estimated that at least 2.5 million people are currently living with cancer in the United Kingdom. This disease is also recognised as the second leading cause

[a]*University of Zululand, Department of Biochemistry and Microbiology, Private Bag X1001, Kwadlangezwa 3886, South Africa. E-mail: DunpallR@unizulu.ac.za*
[b]*University of Zululand, Department of Chemistry, Private Bag X1001, Kwadlangezwa 3886, South Africa. E-mail: RevaprasaduN@unizulu.ac.za*

of death in the United States, contributing to a sum of 1 685 210 new cancer cases and 595 690 cancer deaths annually.[7]

The most predominant treatment approach in cancer involves the use of cytotoxic chemicals that are able to act as antimetabolites, alkylating agents, anti-microtubule agents, topoisomerase inhibitors and cytotoxic antibiotics.[8] These chemotherapeutic agents are administered both orally and intravenously. In addition to combination therapies such as surgery and radioactive therapy, the use of chemotherapy to manage and control the progression of this disease is the most dependable. However these treatments also cause intrinsic toxicity and results in undesirable side effects for the patient. In practise the use of chemotherapy presents several limitations.[9,10] These therapeutic agents lack selectivity and equal biodistribution and are therefore able to attack any cell in the active s-phase of the cell cycle.[11] In addition a large number of these drugs display poor hydrophilicity and have a tendency to aggregate in formulations.[12] These drugs have a low therapeutic index and half-life, which requires higher doses at a greater frequency to ensure therapeutically relevant blood plasma concentrations.[13–15] This results in systemic toxicity, severe side effects such as cancer cachexia syndrome and multidrug resistance. These therapeutic limitations and the overwhelming cancer burden calls for approaches to develop more tailored chemotherapeutic agents.

Nanobiotechnology and nanotherapeutic applications are currently serving as a strategy to improve and advance the delivery of cancer therapeutics.[15] An assortment of nanomaterials such as dendrimers, solid lipid nanoparticles, polymer particles, liposomes, nanoemulsions, metallic and magnetic nanoparticles feature in cancer drug delivery applications.[12] They all display attractive properties that can be specifically manipulated to enhance its biocompatibility, stability and biodegradation. Gold (Au) nanoparticles however has been at the forefront of cancer bioapplications for the past three decades.[11,16] They demonstrate size tunable optical, electronic and catalytic features. During excitation by incident light, gold nanoparticles exhibit surface plasmon resonance (SPR) properties that are characterised by the collective oscillation of valence electrons which results in enhanced light scattering and absorption patterns.[17] These unique optical features enable the rapid conversion of absorbed light into heat resulting in localized photothermal therapy.[18] The intrinsic properties of gold nanoparticles makes them promising components in the assembly and synthesis of gold fabricated core–shell nanostructures that are multifunctional.[19]

The potential to combine gold with a suitable core or shell material that can synergistically function to enhance and improve the efficacy of anti-cancer drugs has gained attention.[20–23] Owing to the inherent properties of biocompatible gold nanoparticles, gold containing core–shell nanoparticles can be designed to increase site specific drug delivery, reduce the systemic toxicity associated side effects, modulate drug release, biodistribution, improve therapeutic efficacy and increase patient compliance whilst utilizing a much lower dose than conventional chemotherapeutic agents.[11,16] The rationale for developing gold core–shell nanoparticles surpass the domains of experimental research, the goal is to develop these nano-drug delivery systems for downstream clinical use. In 1995, the first food and drug administration (FDA) approved nano-drug called Doxil® was introduced, this drug was developed through doxorubicin loading within PEGylated nano-liposomes and has been on the market for the last 20 years.[24] Moreover FDA approved paclitaxel albumin-stabilized nanoparticle formulation under the name Abraxane® has been approved for treatment of breast, non-small cell lung and metastatic pancreatic cancers.[25] Currently there are several gold nano-drug formulations for cancer therapeutics that are either in the clinical phase or FDA approved. It is worth noting that a class of silica core/gold shell nanoparticle has reached the clinical trial phase under the trade name AuroLase.[26] Several FDA approved cancer nanochemotherapeutics are listed in Table 1.

2 Gold based core–shell nanoparticles for cancer drug delivery

During the 1980s composite colloidal semiconductor nanoparticles were first developed, the hypothesis was based on combining two or more compatible materials that would function as a single entity, resulting in superior physiochemical properties over their analogous parent nanoparticles.[27,28] This observation led to the development of heterostructured nanomaterials such as multi-layered semiconductor nanoparticles that were produced with a primary aim of enhancing existing features.[29–32] Core–shell nanoparticles are synthesized by combining two or more different types of nanomaterials. The growth of core–shell nanoparticles is characterised by a distinct core nanomaterial that is surrounded by a layer of shell nanomaterial. Several types of core–shell nanomaterials exist, categories include semiconductor/semiconductor, magnetic/semiconductor and metal/semiconductor heterostructures. Despite the novel properties and bioapplications of gold nanoparticles, in recent decades the demand and need for greater advancements within materials chemistry and bionanotechnology has led to the synthesis of core–shell nanostructures fabricated with either a gold core or gold shell component.[33]

In the chemical assembly of gold core–shell nanoparticles the synthesis usually involves two steps, the synthesis of a core followed by the growth of a shell. These growth techniques can be categorised on the availability of the core material during the synthesis procedure.[34] (I) involves the initial growth of the core followed by the shell particles *in situ* within the same reaction solution. (II) Involves independent growth of the core or seeding materials followed by the addition of the shell material through the use of appropriate coating or surface modifications. In the first technique the core is synthesized using suitable shape directing reactants or surface capping agents, after the core is formed more reactants are added to initiate shell growth *in situ*.[35] In the second technique the core particles are independently synthesized and purified followed by surface modification that facilitates the growth of

Table 1 Clinically approved intravenous cancer nanoparticle medicines. Reprinted in part with permission from Anselmo, A. C. and Mitragotri, S. (2016), Nanoparticles in the clinic, Bioengineering & Translational Medicine, 1: 10–29. doi:10.1002/btm2.10003, © 2016, The Authors. Bioengineering & Translational Medicine is published by Wiley Periodicals, Inc. on behalf of the American Institute of Chemical Engineers.

Name	Particle type/drug	Approved application/indication	Approval (year)	Investigated application/indication	ClinicalTrials.gov identifier
Doxil/Caelyx (Janssen)	Liposomal doxorubicin (PEGylated)	Ovarian cancer (secondary to platinum based therapies) HIV-associated Kaposi's sarcoma (secondary to chemotherapy) Multiple myeloma (secondary)	FDA (1995) EMA (1996)	Various cancers including: solid malignancies, ovarian, breast, leukemia, lymphomas, prostate, metastatic, or liver	166 studies mention Doxil 90 studies mention CAELYX
DaunoXome (Galen)	Liposomal daunorubicin (non-PEGylated)	HIV-associated Kaposi's sarcoma (primary)	FDA (1996)	Various leukemias	32 studies mention DaunoXome
Myocet (Teva UK)	Liposomal doxorubicin (non-PEGylated)	Treatment of metastatic breast cancer (primary)	EMA (2000)	Various cancers including: breast, lymphoma, or ovarian	32 studies mention Myocet
Abraxane (Celgene)	Albumin-particle bound paclitaxel	Advanced nonsmall cell lung cancer (surgery or radiation is not an option) Metastatic breast cancer (secondary) Metastatic pancreatic cancer (primary)	FDA (2005) EMA (2008)	Various cancers including: solid malignancies, breast, lymphomas, bladder, lung, pancreatic, head and neck, prostate, melanoma, or liver	295 studies mention Abraxane
Marqibo (Spectrum)	Liposomal vincristine (non-PEGylated)	Philadelphia chromosome-negative acute lymphoblastic leukemia (tertiary)	FDA (2012)	Various cancers including: lymphoma, brain, leukemia, or melanoma	23 studies mention Marqibo
MEPACT (Millennium)	Liposomal mifamurtide (non-PEGylated)	Treatment for osteosarcoma (primary following surgery)	EMA (2009)	Osteosarcomas	4 studies mention MEPACT: 3 active/recruiting
Onivyde MM-398 (Merrimack)	Liposomal irinotecan (PEGylated)	Metastatic pancreatic cancer (secondary)	FDA (2015)	Various cancers including: solid malignancies, breast, pancreatic, sarcomas, or brain	7 studies mention MM-398/ Onivyde: 6 active/recruiting

the shell material.[36] The effects of temperature, reactant concentration, capping agents, pH and external forces such as sonochemical synthesis and electrodeposition play pivotal roles in maintaining uniform shape and control of shell thickness coating. These features determine the type of applications the gold core–shell nanoparticles may be used in. Establishing proper control over shell thickness and coating is still a challenge.[37] Researchers observe various difficulties in this step such as incomplete or bare core material, aggregation of core material within the reaction solution, formation of alloy materials, formation of shell materials or secondary nanomaterials and precise control of the reaction rates.[33] The use of appropriate surfactants, temperature and pH are employed to change the surface properties of the core materials to make it more suitable for selective deposition and growth of shell particles.[38] Numerous dedicated reviews that cover specific materials composed of Au-based core–shell nanoparticles are reported for SiO_2/Au, Au/TiO_2, Au/polymer, magnetic/Au and Au–Ag core–shell nanoparticles.[39–41] These reports illustrate the diversity of Au-based core–shell nanostructures in scientific research. Gold nanorods/graphene oxide core–shell nanocomposites were synthesized to demonstrate drug targeting capabilities. The use of a graphene oxide shell in combination with gold nanorods played a vital role in decreasing the toxicity of the surfactant covered gold nanorods and facilitated the conjugation of hyaluronic acid.[42] It has been demonstrated that the design of Au containing core–shell nanoparticles can be easily functionalized to accommodate attractive properties suitable for bioapplications other than drug delivery. One of these features include the tuning of optical properties for bio-imaging and light mediated photothermal destruction of target tissues. Near-infrared light-responsive supramolecular nanovalves based on mesoporous silica-coated gold nanorods were developed as a novel cancer theranostic biomaterial. This unique core–shell structure shows the potential to support the uptake and entrapment of a myriad of chemotherapeutic drugs and biomolecular targeting ligands.[20] Appropriate characterisation techniques are used to identify the interaction of the core and shell components, provide insights on the growth mechanisms and other physicochemical features that can be adjusted according to the intended bioapplication.[11,36,42]

The use of pulsed laser ablation of alloy targets in the presence of different solvents was used to prepare and control the phase structure in laser generated iron–gold core–shell nanoparticles. This work highlighted the importance of developing synthesis routes that are able to easily tune and control the thickness of the shell layer thereby altering the functionality of the heterostructure. In Fig. 1, STEM-HAADF elemental mapping characterisation was performed to evaluate the interaction and morphology between the core and shell nanomaterials.[43] The effects of template removal during the synthesis of mesoporous silica-coated gold nanorods is a huge parameter on the morphology and drug delivery applications. The study displayed that structure directing agents such as cetyltrimethylammonium bromide can easily be removed from the Au@SiO_2 core–shell product. Intricate steps in the assembly of the

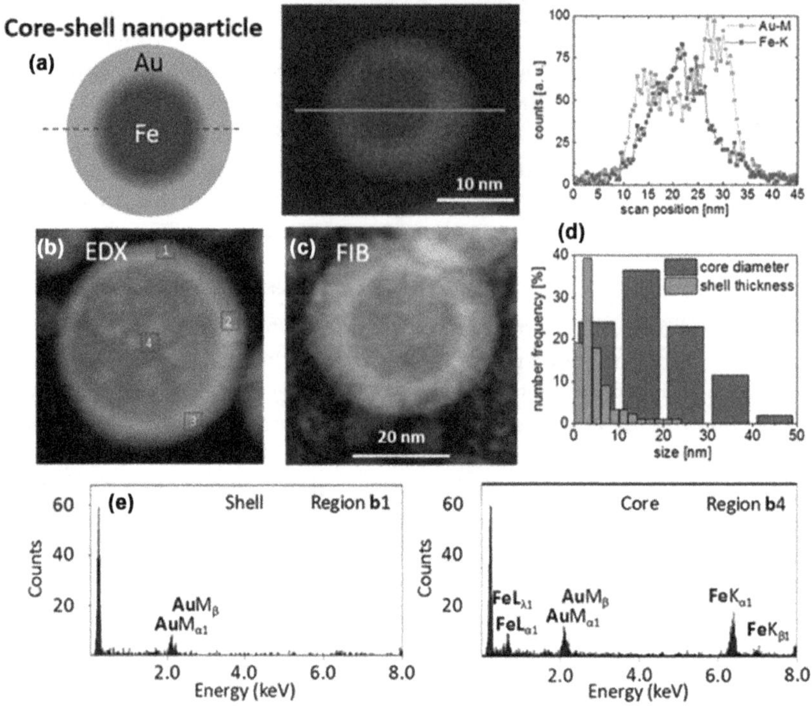

Fig. 1 Analysis of core–shell nanoparticles that were laser-generated in acetone. (a) Line scan and EDX signal of a core–shell nanoparticle. (b) STEM-HAADF image of a Fe@Au core–shell nanoparticle with regions of EDX analysis indicated (shell: regions 1–3, predominantly Au, core: region 4, composed of Fe and Au. (c) STEM–HAADF image of an FIB-milled Fe@Au core–shell nanoparticle. (d) Histograms of the shell thickness and core diameter of laser-generated Fe@Au core–shell nanoparticles. Average values and peak maxima indicate a mean core diameter of 15 nm and a shell thickness of 3 nm. (e) Representative EDX point measurements indicating the presence of pure Au in the shell (region b1) and a mixture of Au and iron in the core (region b4). Reprinted with permission from Wagener *et al.*, Solvent–surface interactions control the phase structure in laser-generated iron–gold core–shell nanoparticles. Scientific reports. 2016; 6. Copyright 2016, Nature Publishing Group. Published under a Creative Commons Attribution 4.0 International License (http://creativecommons.org/licenses/by/4.0/).

core–shell structures have colossal effects of its ability to efficiently load anti-cancer agents within the gold fabricated structure.[44] The stability of the prepared gold fabricated core–shell structures is usually established through temperature and pH adjustments which later play central roles in the conjugation and entrapment of anti-cancer molecules and cellular ligands. The use of carbohydrate, receptor and antibody targeting can be easily applied in the design of gold based core–shell structures. Folic acid-conjugated silica-coated gold nanorods and quantum dots for dual-modality CT and fluorescence imaging and photothermal therapy was prepared with a silane coupling agent. The resultant gold fabricated core–shell structure displayed unique optical properties, low *in vitro* cytotoxicity, improved therapeutic efficacy and higher generated heat capacity.[45]

In an elaborate review, Chaudhuri and Paria reported the synthesis, applications and the progress of research advancements of Au-based core–shell nanoparticles.[36] In summarising the findings of their report, the authors demonstrated that early research on Au-based core–shell nanoparticles focused principally on optimising various synthesis routes, core to shell ratios and core to shell combinations. In optimising the materials chemistry scientists were able to correlate size tunable properties that have contributed to the many advancements within applications of Au-based core–shells. The synthesis routes are broadly classified into two methods *i.e.* wet and dry chemistry.[44,46,47] Wet chemistry routes are further classified as reduction of gold salt, electrochemical synthesis and thermal decomposition and photochemical method. Dry chemistry methods involves the condensation of metal from metal vapour under inert atmospheric conditions commonly known as the laser vaporization/controlled condensation technique. Due to the many advances and improvements in synthetic techniques, it has become possible to produce core–shell nanoparticles in various shapes such as nano disks, nanorods, nanocubes, hexagonal, tubular shapes and nanocage core–shell multi-layered nanomaterials (Fig. 2).[48] The shape of the core–shell, also known as surface anisotropy effects, contributes to the properties of the heterostructured nanomaterials and also influences the potential toxicity linked to the gold fabricated core–shell nanostructure.

In an independent report, following the preparation of Au-nanocage@mSiO_2 core–shell nanocages using a galvanic replacement reaction, the poly-(*N*-isopropyl acrylamide) (PNIPAM) polymer shells were covalently anchored to the surface of the mesoporous silica shells using the atom transfer radical polymerization method.[47] A solution of doxorubicin hydrochloride (Dox) was then loaded within the Au-nanocage@mSiO_2@PNIPAM nanocarrier and used to greatly enhance the *in vitro* drug delivery of doxorubicin using HeLa cells. The combination of

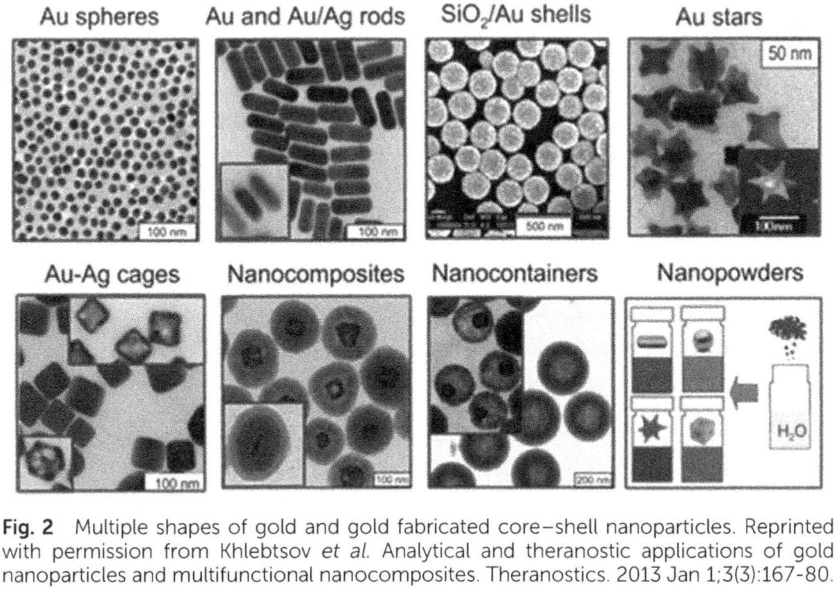

Fig. 2 Multiple shapes of gold and gold fabricated core–shell nanoparticles. Reprinted with permission from Khlebtsov *et al.* Analytical and theranostic applications of gold nanoparticles and multifunctional nanocomposites. Theranostics. 2013 Jan 1;3(3):167-80. Copyright 2016 Ivyspring International Publisher.

gold nanocages and SiO_2 exhibited red shifted optical properties which further contributed to the photothermal destruction of the cancer cells further implicating that the surface plasmon properties are shape-dependent.[47] A major challenge in designing gold fabricated core–shell nanoparticles arises from determining the structure of a two-component system and in identifying compatible core/shell component elements that can each enhance the drug delivery capacity. The optimal use of preparation techniques also contribute to the materials overall stability, biocompatibility and biological application. The core, shell, surface passivating agent and the combination of all three simultaneously serve as drug carriers that are stable in physiological environment, displays high drug loading capacity and low toxicity.

The controlled synthesis of bimetallic heterosystems such as Pd/Pt or Ag/Au core/shell systems is possible because of the relatively close match between the crystal lattices of their component parts. In contrast, core/shell metal–semiconductor systems are more difficult to synthesize due to their differing crystal structures and large crystal lattice mismatches.[49,50] There have been few reports of metal–semiconductor core/shell nanostructures.[51,52] Wang and co-workers described a general method for the preparation of Au nanocrystals of different shapes and sizes as starting material for the synthesis of water-dispersible Au–metal sulfide core/shell structures.[53] Salant *et al.* conducted elaborate work on the growth of gold tips on anisotropic CdSe particles which was used to establish a relationship between the growth pattern and morphology that develops from a stabilized core.[46] The synthesis of gold shelled heterostructures with the core composed of a magnetic nanomaterial such as iron oxide is well reported.[54] The gold shell reduces the toxicity of the iron oxide and also enhances the biocompatibility properties.[54] When gold is used as the core component it exhibits certain advantages; playing a direct role in the size tunability and morphology of the core/shell or heterostructured nanoparticle. The advantages of using gold as a core or shell component depends also on its chemical interaction with other materials therefore the selection of chemically compatible materials is important in identifying how the gold fabricated core–shell will interact with the chemotherapeutic drug.[55] In general metal/semiconductor hybrid nanostructures are very interesting as they possess strong interactions between the metal plasmon and the semiconductor exciton *i.e.* electron/hole pair.[56] The recent synthesis of Au–ZnTe core–shell nanoparticles has demonstrated that the shape, size and thickness of the shell material directly impacts the toxicity, bio-conjugation capability and optical properties, indicating that in merely changing the shape from spheres to rods the surface plasmon resonance properties can be greatly tuned.[57] Au–ZnTe nanoparticles displayed unique growth patterns that directed the thickness of the ZnTe shell layer. In tuning the shape, size and surface features of the gold core, the ZnTe shell thickness could be greatly increased, thereby improving its bio-imaging and drug conjugation capacity.[57,58]

In a similar report ZnO–Au core–shell nanoparticles was prepared for tumour targeting applications using camptothecin. The results indicated that the conjugated camptothecin was released through a diffusion-controlled mechanism. The strong hydrophobic interactions of the hydrophobic poly (l-lactide) with the drug molecules could be a main reason for the slow and steady release of the chemotherapeutic drug.[59] To integrate multiple applications in a single gold fabricated nanocomposite mesoporous silica-coated gold nanorod was developed. The compatible system was designed to utilize the optical properties of the gold core for hyperthermal activity whilst the mesoporous silica shell was used as the reservoir of photosensitizer (Al(III) phthalocyanine chloride tetrasulfonic acid, $AlPcS_4$). Interestingly the tumour targeting ligand lactobionic acid and long-circulating poly (ethylene glycol) chains were introduced *via in vitro* interactions with hepatoma cells.[60] The structural diversity of gold fabricated core–shell structures are vast, most common shapes include the design of nanorods, nanostars and nanocages. The assembly of nanocages shows promise in developing nano carriers that display high entrapment and controlled release efficiencies.[47] In work reported by Zhang *et al.*, dox conjugated Au@SiO_2 nanorods were synthesised using thermo- and pH-responsive polymer shell called poly (*N*-isopropylacrylamide-*co*-acrylic acid). The gold fabricated nanorods displayed red shifted optical wavelengths in the range of 700–900 nm. The characterisation techniques used in this study is shown in Fig. 3. The analysis further demonstrated the effect of pH on the cumulative release of the loaded drug.[61] At an acidic pH the drug release had dramatically

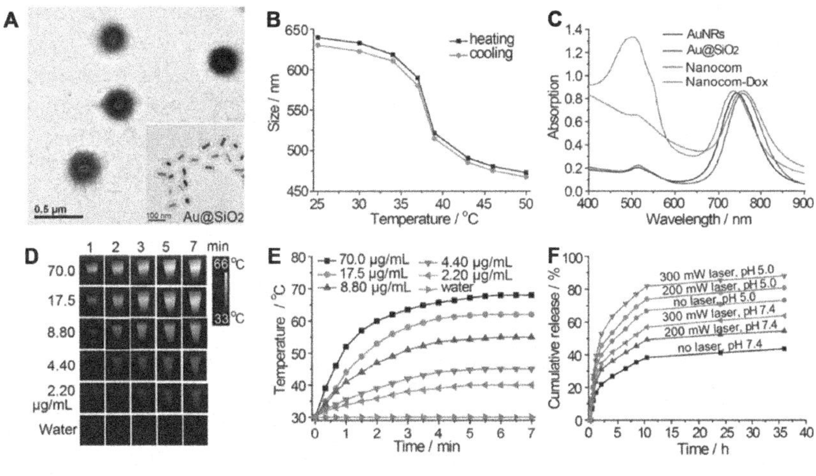

Fig. 3 Characterization of the nanocompostie and laser-controlled drug release profiles. (A) TEM images of the nanocomposite and Au@SiO_2 (inset). (B) Reversible hydrodynamic size change of the nanocomposite with respect to temperatures. (C) Extinction spectra of AuNRs, Au@SiO_2, the nanocomposite (abbreviated as Nanocom in all figures), and Dox-loaded nanocomposite (abbreviated as Nanocom-Dox in all figures). (D) The temperature images of nanocomposite suspension under NIR laser (760 nm, 500 mW, 16 W cm^{-2}) irradiation were recorded by a IR camera at different concentration. (E) The rate of temperature rise and the final temperature were proportional to particle concentration at the constant laser function power (760 nm, 500 mW, 16 W cm^{-2}). (F) Dox release profiles from Nanocom-Dox with or without NIR laser irradiation at different pHs. Reprinted with permission from Zhang *et al.* Near infrared laser-induced targeted cancer therapy using thermoresponsive polymer encapsulated gold nanorods. Journal of the American Chemical Society. 2014 May 9;136(20):7317-26. Copyright 2014 American Chemical Society.

increased, these findings could be deemed rather useful in clinical applications of active tumours that support a more acidic extracellular matrix.[61]

3 Nano-drug formulations for cancer drug delivery

The distinct features of an actively growing tumour provides vital information that aid in improving the design of cancer targeting nanotherapeutics and is linked to the development of stimuli responsive core–shell nanoparticles that release the drug load upon various internal and external triggers.[62] Nano-drug formulations can be designed to target a tumour site through passive, active or both mechanisms simultaneously (Fig. 4).[63] The passive targeting mechanism is based on a property known as the enhanced permeability and retention effect, which causes the presence of leaky vasculature anatomy that permits the entry and accumulation of nano-drug formulations in higher concentrations. The active targeting mechanism involves the use of cellular ligands and directing molecules such as proteins, glucose, folates, antibodies or hormones that bind specifically to the overexpressed receptors present on the cancer cell membrane.[62–64] Active binding mechanisms involve the attachment of ligand molecules to the surface of the nano-drug. Once the nano-drug is exposed to the tumour cells the ligand attaches to the membrane bound receptor and initiates cellular uptake through a process called receptor mediated endocytosis causing the nano-drug to accumulate within the tumour cells.[12] Overall the poor vasculature, ineffective lymphatic drainage and lack of oxygen within the growing tumour tissues causes a state of hypoxia which contributes to the acidic microenvironment around cancer cells. The subtle changes in pH, physiological temperature, inflammatory markers, reactive oxygen species, hormone and enzyme levels are important triggers that differentiates normal and cancerous tissues.[65]

The design and application of nanotherapeutics that can respond to redox-sensitivity is gaining huge interest in the pharmaceutical domain of drug development. The reduction of glutathione (GSH) is an eminent redox reaction within cancer cells.[66] The level of GSH present in cancer cells are at least 500 times higher when compared to normal cells.[67] Tumour targeting gold nanostars conjugated to cyclic arginine–glycine–aspartic acid (RGD) peptides and doxorubicin was covalently linked to the surface of Au nanostars *via* glutathione (GSH) to form Au–DOX, Au–cRGD, and Au–cRGD–DOX.[68] The multifunctional Au–DOX, Au–cRGD, and Au–cRGD–DOX was used to monitor drug delivery and photothermal therapies. The findings from this work confirmed that the Au nanostars entered and interacted with the intracellular environment indicative through the red fluorescence observed from the nucleus, confirming that free doxorubicin was released from Au–cRGD–DOX as a result of the replacement of GSH on Au NS surface by the abundant GSH present within the tumour cell.[68] Other endogenous triggers such as changes in pH levels within cancer cells are well reported and used to assist the gold core–shell nanoparticles to release its potent drug load

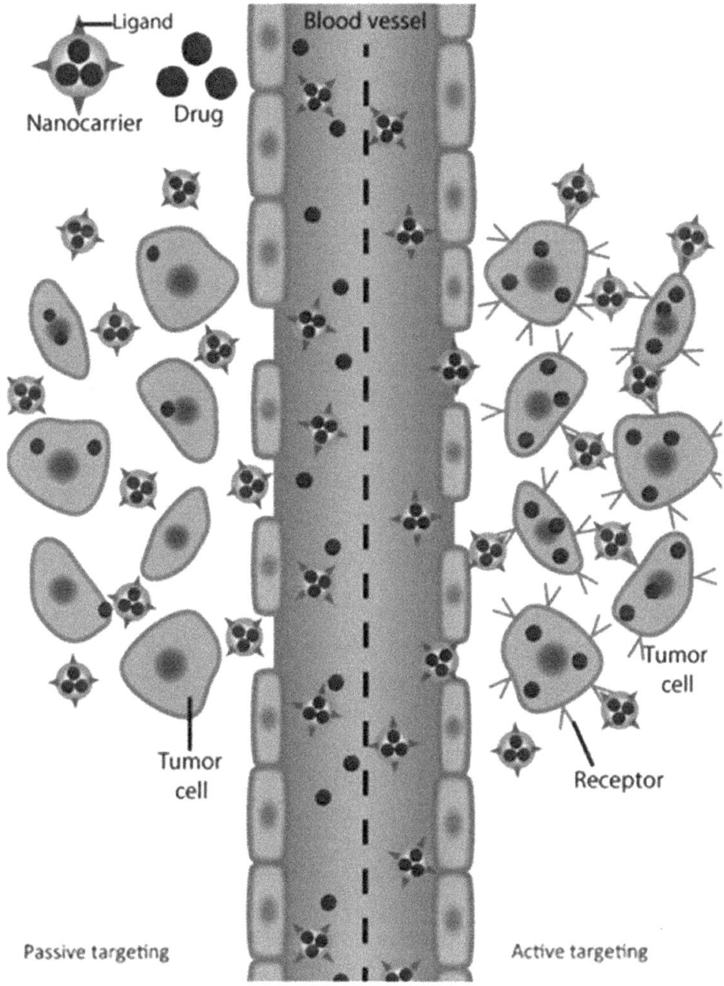

Fig. 4 Graphic illustration of the accumulation of gold nanoparticles delivering chemotherapeutic agents into the tumour through passive or active targeting mechanisms. Reprinted with permission from Journal of Experimental & Clinical Medicine, Volume 6, G. Ajnai, A. Chiu, T. Kan, C.-C. Cheng, T.-H. Tsai and J. Chang J, Trends of gold nanoparticle-based drug delivery system in cancer therapy, 172-8, Copyright 2014, with permission from Elsevier.

only once the acidic environment is detected.[69] The level of pH is an extremely vital aspect in maintaining homeostasis within the human body. The pH level is relative to a well-designed function which differs considerably for various organs, cancer and normal tissues. For example the pH of the stomach is greatly acidic when compared to the pH of the intestinal tract.[70] Nanotherapeutics have to be designed accordingly to ensure that pH of the target site can be used as a trigger to release the drugs. The use of ionizable compounds, such as carboxylic, phosphoric, sulfonic and amino acids can be conjugated to the nanotherapeutic to facilitate pH responsive drug targeting.[12,71,72] The nano-drug formulation responds to pH changes by undergoing an initial dissociation phase which causes the nano-drug formulation to destabilize through swelling or collapsing. This creates fluctuations in the partition coefficient between the entrapped or conjugated drug and nanoparticle. Another strategy for pH-sensitive drug release is to use acid-labile linkers to covalently conjugate the inactive prodrug to the surfaces of nanotherapeutics, allowing the drug to only be release once the linker is hydrolysed.[70]

Recently the synthesis of multi-walled carbon nanotubes (MWCNTs) decorated with gold/silver core–shell nanoparticles (Au@Ag NPs) and fluorescein isothiocyanate for tracking the intracellular doxorubicin drug release triggered by pH variation was reported.[17] The study demonstrated a detailed release process and carrier distribution that was recorded as environmental pH changes during cell endocytosis. Increased anticancer drug therapeutic efficacy was also achieved using graphene oxide and AuNP core polyaniline shell (GO–Au@PANI) nanocomposites. The increased *in vitro* drug release was demonstrated in lower pH values from phosphate buffer solutions using fluorescence spectroscopy measurements for detection. Although the use of pH responsive drug release in nanotherapeutics is widely reported many studies indicate that the combination of additional stimuli such as temperature, redox or magnetic parameters could greatly increase the rate and control of drug release within the tumour environment.[17]

Physiological temperature is one of the most effective parameters used to control drug release in nanotherapeutic cancer drug delivery applications.[73] The metabolic demands of a rapidly growing tumour exceeds the temperature of 37 °C. Gold fabricated core–shell nanoparticles can be designed to remain stable at physiological temperatures and release the drug load at temperatures between 40–45 °C.[4] This feature allows the nanotherapeutic to selectively attack and destroy the tumour tissues. In a single study both photothermal and temperature responsive properties were demonstrated using Au@SiO_2 core–shell nanoparticle. The nanorods structure was further functionalized with poly (*N*-isopropylacrylamide-*co*-acrylic acid) and drug loaded with dox.[61] The activation stimulus in this study was transformed from heat to a NIR laser, which displayed a direct influence on both size and drug release. Interestingly the thermoresponsive polymer shell aided in prolonging the composites blood circulation time. Exogenous triggers such as the presence of a magnetic field, ultrasound (US), light, electric pulsation and high energy radiation also have the potential to activate drug release using external technologies.[74] Photo sensitive nanotherapeutics show the ability to regulate drug release through stimulation by light irradiation.[70] However, the light penetration depth greatly limits the non-invasive drug delivery for tumours located within deeper tissues or internal organs. The use of magnetic stimuli may deliver a non-invasive approach to spatially regulate the drug targeted features through controlled exposure of an external magnetic field.

A multifunctional oligonucleotide-gated silica shell-coated Fe_3O_4–Au core–shell system was designed to use magnetism as an external trigger, by remotely activating drug release. The results verified the significant therapeutic effects on tumours with the assistance of combination therapy consisting of magnetic guidance and remote NIR control which enhanced tumour destruction.[15,73]

4 Pharmaceutical formulation and drug development

The pharmacokinetic and pharmacodynamics properties of drug formulations can be largely altered using nanomaterials.[74] A key feature in developing successful gold fabricated core–shell nano-drug carriers is based on two important fundamentals;

I. Physiochemical features such as size, shape, composition and other physicochemical properties of the nanomaterial. Additionally, depending on the preparation methods biomolecules and chemotherapeutic drugs can be dissolved, entrapped, encapsulated or surface conjugated. It is therefore critical to establish a compatible core–shell system with superior biocompatibility and stability features.[39,66,75]

II. Physicochemical properties of the chemotherapeutic drug is essential in understanding how the drug will interact with the nano carrier in terms of its drug loading or surface conjugation capacity and also its subsequent therapeutic efficacy. Most chemotherapeutic agents can be categorised according to their aqueous solubility as either hydrophobic or hydrophilic. In addition to hydrophilicity, the chemotherapeutic chemicals can be further categorised by the electrostatic characteristics as either highly charged or neutral.[12]

The ability to effectively deliver hydrophobic drugs that generally tend to aggregate in formulation is a major challenge. Suitable nanoformulations provide a valuable platform to enhance the delivery and cellular uptake of hydrophobic chemotherapeutic agents if they are well encapsulated with hydrophilic carriers.[15,63] A thiol-terminated polyethylene glycol (PEG400)–paclitaxel conjugated gold nanoparticle core surrounded by a liposome shell structure was employed to efficiently deliver paclitaxel to liver cancer cells. The study clearly indicated superior drug loading and stability features in the presence of a liposome shell.[76] Hydrophilic chemotherapeutic agents offer additional challenges, some of these drugs display poor miscibility in formulations, are unable to the bypass the lipid rich cell membrane channels and they exhibit low half-life within systemic circulation. The use of prodrug formulations that are only active upon multiple stimuli such as enzyme, temperature and pH is quite common. A biocompatible protein–gold nano construct (AFt–Au) with a core–void–shell structure containing 5-fluorouracil (5-FU) was designed to exhibits a high sensitivity towards carcinoma cells. In this study the conjugated 5-FU was internalized into HepG2 cancer cells with AFt–Au primarily via receptor-mediated endocytosis. The design of AFt–Au, was used as a bio-nano-chemosensitizer and increased the cells susceptibility to 5-FU through cell-cycle regulation mechanisms.[77] In another study the absorption of anticancer curcumin extract was improved through a formulation of hydroxypropyl-beta-cyclodextrin

(HP-β-CD)-modified GoldMag nanoparticles (CD-GMNs). The curcumin loading, encapsulation efficiency and releasing properties were dramatically enhanced showing a drug encapsulation capacity of 88%. Curcumin release was activated through pH sensitivity, showing huge promise not only for conventional chemotherapeutics but natural alternative medicines as well.[75] Chemotherapeutic agents are formulated into gold based core–shell nanoparticles either through external surface intermolecular binding mechanisms such as covalent, noncovalent and chemisorption or internal loading or encapsulation within the nano core material. The gold fabricated core–shell nano-drug carriers must be designed to easily support both passive and active drug targeting mechanisms for cancer drug delivery. The drug entrapment efficiency is determined using ultraviolet–visible spectroscopy and high performance liquid chromatography techniques.[75] The collected data in then extracted into the following equation to determine how much of drug is encapsulated;[76]

$$\text{Entrapment efficiency (\%)} = \frac{\text{Total amount of the drug} - \text{amount of the free drug}}{\text{Total drug}} \times 100$$

The therapeutic efficacy of most nano-drug systems is largely related to the entrapment efficiency, stability of the formulation and the mechanism of drug release. Various characterisation techniques are used to monitor the entrapment efficiency and stability of nanotherapeutics. These parameters include changes to particle size, surface charge and particle distribution. The use of UV–Vis spectrum measurements was employed to monitor the drug loading of docetaxel (DTX).[78] The design of an organic–inorganic hybrid nanocomposite was synthesized by deposition of Au onto the surface of docetaxel (DTX)-loaded poly (lactide-*co*-glycolide) nanoparticle cores to form the core–shell nanotherapeutic formulation. Additionally targeting peptide, angiopep-2, was bioconjugated to the gold nanoshell through Au–S bond, achieving drug delivery with active targeting capability. Again the nanotherapeutic formulation showed superior therapeutic efficacy over the conventional docetaxel control.[79] The surface binding of chemotherapeutic agents using gold fabricated core–shell nanoformulations relies on the accurate control over intermolecular interactions that are able to produce stabilised formulations that can easily penetrate the cytosolic region of the cell, drug release is then triggered by the lysosomal degradation that occurs within vacuoles.[78] The use of tumour targeting tLyP-1 peptide ligand, was conjugated to AuNR@SiO_2 nanoparticles and subsequently loaded with anticancer drug camptothecin. AuNR@SiO_2–tLyP-1 showed increased cytotoxicity in Hela and MCF-7 cells. The formulation of AuNR@SiO_2–tLyP-1 displayed unique cellular uptake and drug release that was facilitated through the heat generating properties of Au–SiO_2 which caused instantaneous apoptosis of the cancer cell targets through photothermal destruction.[80] The ability to control drug release, bioavailability and systemic stability still remains a critical challenge in the clinical translation of many gold fabricated core–shell nanoparticles.[64]

5 Future perspectives for gold based core–shell nanoparticles

The use of gold fabricated core–shell nanoparticles are certainly a class of nanotherapeutics that are worth further investigating. The opportunities presented for scale up and commercialisation are largely based on the clinical performance of these nanotherapeutics. Other limiting factors include concerns over dispersity, aggregated growth, biodegradation and bioaccumulation of the nano-vehicle after it has released its cytotoxic drug cargo. The precise control of drug targeting and selectivity of cancers at different stages still remains a challenge especially for personalised chemotherapeutic medicine. Future perspectives for the design and formulation of gold based core–shell heterostructures will involve the use of multiple bioconjugates both internally and externally giving rise to nanostructures that exceed a drug delivery capacity by several folds.

Acknowledgements

The authors wish to acknowledge the University of Zululand, Department of Chemistry, Nanotechnology Research Group and the Department of Science and Technology (DST) and National Research Foundation (NRF) of South Africa through the DST/NRF South African Research Chairs Initiative (SARChi) program.

References

1 D. Hanahan and R. A. Weinberg, *Cell*, 2000, **100**, 57–70.
2 F. Colotta, P. Allavena, A. Sica, C. Garlanda and A. Mantovani, *Carcinogenesis*, 2009, **30**, 1073–1081.
3 D. Hanahan and R. A. Weinberg, *Cell*, 2011, **144**, 646–674.
4 N. J. Krogan, S. Lippman, D. A. Agard, A. Ashworth and T. Ideker, *Mol. Cell*, 2015, **58**, 690–698.
5 O. Warburg, *Science*, 1956, **123**, 309–314.
6 R. L. Siegel, K. D. Miller and A. Jemal, *Ca-Cancer J. Clin.*, 2015, **65**, 5–29.
7 R. L. Siegel, K. D. Miller and A. Jemal, *Ca-Cancer J. Clin.*, 2016, **66**, 7–30.
8 V. T. DeVita and E. Chu, *Cancer Res.*, 2008, **68**, 8643–8653.
9 R. Siegel, C. DeSantis, K. Virgo, K. Stein, A. Mariotto, T. Smith, D. Cooper, T. Gansler, C. Lerro and S. Fedewa, *Ca-Cancer J. Clin.*, 2012, **62**, 220–241.
10 C. E. DeSantis, C. C. Lin, A. B. Mariotto, R. L. Siegel, K. D. Stein, J. L. Kramer, R. Alteri, A. S. Robbins and A. Jemal, *Ca-Cancer J. Clin.*, 2014, **64**, 252–271.
11 A. Kumar, X. Zhang and X.-J. Liang, *Biotechnol. Adv.*, 2013, **31**, 593–606.
12 T. Sun, Y. S. Zhang, B. Pang, D. C. Hyun, M. Yang and Y. Xia, *Angew. Chem., Int. Ed.*, 2014, **53**, 12320–12364.
13 S. Bamrungsap, Z. Zhao, T. Chen, L. Wang, C. Li, T. Fu and W. Tan, *Nanomedicine*, 2012, **7**, 1253–1271.
14 S. D. Steichen, M. Caldorera-Moore and N. A. Peppas, *Eur. J. Pharm. Sci.*, 2013, **48**, 416–427.

15 S. Dhanasekaran, *Int. J. Immunother. Cancer Res.*, 2015, **1**(1), 008–012, **8591**, 1–100.
16 S. Jain, D. Hirst and J. O'Sullivan, *Br. J. Radiol.*, 2012, **85**(1010), 101–113.
17 H. Chen, Z. Liu, S. Li, C. Su, X. Qiu, H. Zhong and Z. Guo, *Theranostics*, 2016, **6**, 1096.
18 L. Vigderman and E. R. Zubarev, *Adv. Drug Delivery Rev.*, 2013, **65**, 663–676.
19 W. Liu, Z. Zhu, K. Deng, Z. Li, Y. Zhou, H. Qiu, Y. Gao, S. Che and Z. Tang, *J. Am. Chem. Soc.*, 2013, **135**, 9659–9664.
20 H. Li, L.-L. Tan, P. Jia, Q.-L. Li, Y.-L. Sun, J. Zhang, Y.-Q. Ning, J. Yu and Y.-W. Yang, *Chem. Sci.*, 2014, **5**, 2804–2808.
21 W.-P. Li, P.-Y. Liao, C.-H. Su and C.-S. Yeh, *J. Am. Chem. Soc.*, 2014, **136**, 10062–10075.
22 S. Shen, H. Tang, X. Zhang, J. Ren, Z. Pang, D. Wang, H. Gao, Y. Qian, X. Jiang and W. Yang, *Biomaterials*, 2013, **34**, 3150–3158.
23 A. S. Monem, N. Elbialy and N. Mohamed, *Int. J. Pharm.*, 2014, **470**, 1–7.
24 Y. C. Barenholz, *J. Controlled Release*, 2012, **160**, 117–134.
25 V. Roy, B. LaPlant, G. Gross, C. Bane, F. Palmieri and N. C. C. T. Group, *Ann. Oncol.*, 2008, mdn661.
26 J. Li, S. Gupta and C. Li, *Quant. Imaging Med. Surg.*, 2013, **3**, 284–291.
27 G. E. Gillberg-LaForce and G. Khanarian, *Journal*, US Pat. 4,828,758. 1989 May 9.
28 K. Takita, K. Masuda, H. Kudo and S. Seki, *Appl. Phys. Lett.*, 1980, **37**, 460–462.
29 R. W. Henkens, J. Zhao and J. P. O'daly, *Journal*, US Pat. 5,225,064. 1993 Jul 6.
30 T. Valla, P. Pervan and M. Milun, *Surf. Sci.*, 1994, **307**, 576–581.
31 C. A. Foss Jr., G. L. Hornyak, J. A. Stockert and C. R. Martin, *J. Phys. Chem.*, 1994, **98**, 2963–2971.
32 G. L. Hornyak, C. J. Patrissi and C. R. Martin, *J. Phys. Chem. B*, 1997, **101**, 1548–1555.
33 R. Ghosh Chaudhuri and S. Paria, *Chem. Rev.*, 2011, **112**, 2373–2433.
34 M. S. Nashner, A. I. Frenkel, D. Somerville, C. W. Hills, J. R. Shapley and R. G. Nuzzo, *J. Am. Chem. Soc.*, 1998, **120**, 8093–8101.
35 R. T. Tom, A. S. Nair, N. Singh, M. Aslam, C. Nagendra, R. Philip, K. Vijayamohanan and T. Pradeep, *Langmuir*, 2003, **19**, 3439–3445.
36 R. G. Chaudhuri and S. Paria, *Kirk-Othmer Encycl. Chem. Technol.*, 2013, DOI: 10.1002/0471238961.goldpari.a01.
37 C. D. Jones and L. A. Lyon, *Langmuir*, 2003, **19**, 4544–4547.
38 P. Reiss, M. Protiere and L. Li, *Small*, 2009, **5**, 154–168.
39 M. Zhao, K. Deng, L. He, Y. Liu, G. Li, H. Zhao and Z. Tang, *J. Am. Chem. Soc.*, 2014, **136**, 1738–1741.
40 H. Yang, J. Shao and B. Q. Li, InASME 2014 International Mechanical Engineering Congress and Exposition 2014 Nov 14 (pp. V010T13A033–V010T13A033). American Society of Mechanical Engineers.
41 A. Tanaka, K. Fuku, T. Nishi, K. Hashimoto and H. Kominami, *J. Phys. Chem. C*, 2013, **117**, 16983–16989.
42 C. Xu, D. Yang, L. Mei, Q. Li, H. Zhu and T. Wang, *ACS Appl. Mater. Interfaces*, 2013, **5**, 12911–12920.
43 P. Wagener, J. Jakobi, C. Rehbock, V. S. K. Chakravadhanula, C. Thede, U. Wiedwald, M. Bartsch, L. Kienle and S. Barcikowski, *Sci. Rep.*, 2016, **6**, 23352.
44 J. Feng, Z. Wang, B. Shen, L. Zhang, X. Yang and N. He, *RSC Adv.*, 2014, **4**, 28683–28690.

45 H.-X. Xia, X.-Q. Yang, J.-T. Song, J. Chen, M.-Z. Zhang, D.-M. Yan, L. Zhang, M.-Y. Qin, L.-Y. Bai and Y.-D. Zhao, *J. Mater. Chem. B*, 2014, **2**, 1945–1953.
46 A. Salant, E. Amitay-Sadovsky and U. Banin, *J. Am. Chem. Soc.*, 2006, **128**, 10006–10007.
47 J. Yang, D. Shen, L. Zhou, W. Li, X. Li, C. Yao, R. Wang, A. M. El-Toni, F. Zhang and D. Zhao, *Chem. Mater.*, 2013, **25**, 3030–3037.
48 N. Khlebtsov, V. Bogatyrev, L. Dykman, B. Khlebtsov, S. Staroverov, A. Shirokov, L. Matora, V. Khanadeev, T. Pylaev and N. Tsyganova, *Theranostics*, 2013, **3**, 167–180.
49 T. J. Slater, A. Macedo, S. L. Schroeder, M. G. Burke, P. O'Brien, P. H. Camargo and S. J. Haigh, *Nano Lett.*, 2014, **14**, 1921–1926.
50 J.-S. Lee, E. V. Shevchenko and D. V. Talapin, *J. Am. Chem. Soc.*, 2008, **130**, 9673–9675.
51 J. Li and H. C. Zeng, *Angew. Chem.*, 2005, **117**, 4416–4419.
52 E. V. Shevchenko, M. I. Bodnarchuk, M. V. Kovalenko, D. V. Talapin, R. K. Smith, S. Aloni, W. Heiss and A. P. Alivisatos, *Adv. Mater.*, 2008, **20**, 4323–4329.
53 Z. Sun, Z. Yang, J. Zhou, M. H. Yeung, W. Ni, H. Wu and J. Wang, *Angew. Chem., Int. Ed.*, 2009, **48**, 2881–2885.
54 I. Y. Goon, L. M. Lai, M. Lim, P. Munroe, J. J. Gooding and R. Amal, *Chem. Mater.*, 2009, **21**, 673–681.
55 E. Jaeger, A. Jaeger, P. Chytil, T. Etrych, B. Říhová, F. C. Giacomelli, P. Štěpánek and K. Ulbrich, *J. Controlled Release*, 2013, **165**, 153–161.
56 S. Brovelli, R. D. Schaller, S. Crooker, F. García-Santamaría, Y. Chen, R. Viswanatha, J. A. Hollingsworth, H. Htoon and V. I. Klimov, *Nat. Commun.*, 2011, **2**, 280.
57 R. Dunpall, E. A. Lewis, S. J. Haigh, P. O'Brien and N. Revaprasadu, *J. Mater. Chem. B*, 2015, **3**, 2826–2833.
58 R. Dunpall, S. Mlowe and N. Revaprasadu, *New J. Chem.*, 2014, **38**, 6002–6007.
59 T. Chen, T. Zhao, D. Wei, Y. Wei, Y. Li and H. Zhang, *Carbohydr. Polym.*, 2013, **92**, 1124–1132.
60 G. F. Luo, W. H. Chen, Q. Lei, W. X. Qiu, Y. X. Liu, Y. J. Cheng and X. Z. Zhang, *Adv. Funct. Mater.*, 2016, **26**(24), 4339–4350.
61 Z. Zhang, J. Wang, X. Nie, T. Wen, Y. Ji, X. Wu, Y. Zhao and C. Chen, *J. Am. Chem. Soc.*, 2014, **136**, 7317–7326.
62 R. Sinha, G. J. Kim, S. Nie and D. M. Shin, *Mol. Cancer Ther.*, 2006, **5**, 1909–1917.
63 G. Ajnai, A. Chiu, T. Kan, C.-C. Cheng, T.-H. Tsai and J. Chang, *J. Exp. Clin. Med.*, 2014, **6**, 172–178.
64 K. Chatterjee, S. Sarkar, K. J. Rao and S. Paria, *Adv. Colloid Interface Sci.*, 2014, **209**, 8–39.
65 I. I. Lungu, M. Rădulescu, G. D. Mogoşanu and A. M. Grumezescu, *Rom. J. Morphol. Embryol.*, 2016, **57**, 23–32.
66 H. Zheng, L. Yin, X. Zhang, H. Zhang, R. Hu, Y. Yin, T. Qiu, X. Xiong and Q. Wang, *J. Biomed. Nanotechnol.*, 2016, **12**, 1641–1653.
67 W.-H. Chen, G.-F. Luo, Q. Lei, F.-Y. Cao, J.-X. Fan, W.-X. Qiu, H.-Z. Jia, S. Hong, F. Fang and X. Zeng, *Biomaterials*, 2016, **76**, 87–101.
68 H. Chen, X. Zhang, S. Dai, Y. Ma, S. Cui, S. Achilefu and Y. Gu, *Theranostics*, 2013, **3**(9), 633–649.
69 J. Liu, Y. Huang, A. Kumar, A. Tan, S. Jin, A. Mozhi and X.-J. Liang, *Biotechnol. Adv.*, 2014, **32**, 693–710.
70 S. Mura, J. Nicolas and P. Couvreur, *Nat. Mater.*, 2013, **12**, 991–1003.
71 P. Chen, Z. Wang, S. Zong, D. Zhu, H. Chen, Y. Zhang, L. Wu and Y. Cui, *Biosens. Bioelectron.*, 2016, **75**, 446–451.

72 M. Ghorbani, H. Hamishehkar, N. Arsalani and A. A. Entezami, *J. Nanopart. Res.*, 2015, **17**, 1–13.
73 A. C. Anselmo and S. Mitragotri, *Bioeng. Transl. Med.*, 2016, **1**, 10–29.
74 D. Liu, F. Yang, F. Xiong and N. Gu, *Theranostics*, 2016, **6**(9), 1306–1323.
75 T. Lian, M. Peng, A. J. Vermorken, Y. Jin, Z. Luo, W. J. Van de Ven, Y. Wan, P. Hou and Y. Cui, *J. Nanosci. Nanotechnol.*, 2016, **16**, 6258–6264.
76 Q.-Y. Bao, N. Zhang, D.-D. Geng, J.-W. Xue, M. Merritt, C. Zhang and Y. Ding, *Int. J. Pharm.*, 2014, **477**, 408–415.
77 X. Liu, W. Wei, S. Huang, S.-S. Lin, X. Zhang, C. Zhang, Y. Du, G. Ma, M. Li and S. Mann, *J. Mater. Chem. B*, 2013, **1**, 3136–3143.
78 M. Chanana, P. Rivera_Gil, M. A. Correa-Duarte, L. M. Liz-Marzán and W. J. Parak, *Angew. Chem., Int. Ed.*, 2013, **52**, 4179–4183.
79 Y. Hao, B. Zhang, C. Zheng, R. Ji, X. Ren, F. Guo, S. Sun, J. Shi, H. Zhang and Z. Zhang, *J. Controlled Release*, 2015, **220**, 545–555.
80 B. Xu, Y. Ju, Y. Cui, G. Song, Y. Iwase, A. Hosoi and Y. Morita, *Langmuir*, 2014, **30**, 7789–7797.

Supramolecular chemistry of AIE-active tetraphenylethylene luminophores

Duong Duc La,[a] Anuradha,[a] Akhil Gupta,[b] Mohammad Al Kobaisi,[a] Anushri Rananaware[a] and Sheshanath V. Bhosale*[a]

DOI: 10.1039/9781782620358-00075

This book chapter provides recent progress in the development of chemistry of aggregation-induced emission (AIE)-active tetraphenylethylene luminophores, their synthetic strategies, and their applications in the research fields of optoelectronics, sensing and supramolecular science comprehensively. It begins with a discussion on a collection of synthetic structural motifs. We will also explore novel derivatives, structures, electronic and spectroscopic properties of the TPE luminophores. The practical applications include certain topics in medicinal, sensing, optoelectronic devices and supramolecular architectures. We have illustrated advances in tetraphenylethylene research in the mechanoluminescent materials followed by exploring recent development of functionalization/decoration of AIE-active tetraphenylethylene luminogens onto an aggregation-caused quenching (ACQ) fluorophores, such as PDI, NDI and porphyrins, converting the whole chromophore into AIE luminogens with possible applications.

1 Introduction

Luminescent materials with aggregation-induced emission (AIE) properties are of both scientific and technological interest. This is important because the luminophores are commonly used in film or particulate forms, for example, films of organic light emitting diodes (OLED), and particular aggregate with a high contrast in chemo- and biosensors.[1,2] Since the discovery of AIE concept in 2001 with abnormal behaviours resulting from it, considerable research intention has been paid to this concept.[3] The AIE phenomena is opposite to the common aggregation-caused quenching (ACQ) encountered when most conventional luminophores/fluorophores aggregate.[4] AIE-effect becomes the main molecular exited state relaxation pathway when other means of relaxation are restricted especially relaxation *via* intramolecular rotation (RIR).[5,6] The literature study shows that there have been various chemical, physical, and engineering approaches that were taken to explore luminophore aggregation, for example attachment of bulky alicyclics, encapsulation by amphiphilic, and blending with polymers, to name a few.[7,8] However, effects such as the steric effects of bulky alicyclics can twist the conformations of the chromophoric units and hinder the charge transport in electroluminescence (EL) devices in luminophores π-conjugation.

Among AIE-active luminophores tetraphenylethylene (TPE), hexaphenylsilole (HPS) and distyreneanthracene (DSA) (see Scheme 1)

[a]*School of Science, RMIT University, GPO Box 2476, Melbourne, VIC-3001, Australia. E-mail: sheshanath.bhosale@rmit.edu.au; bsheshanath@gmail.com*
[b]*Institute for Frontier Materials, Deakin University, Waurn Ponds, Victoria 3216, Australia*

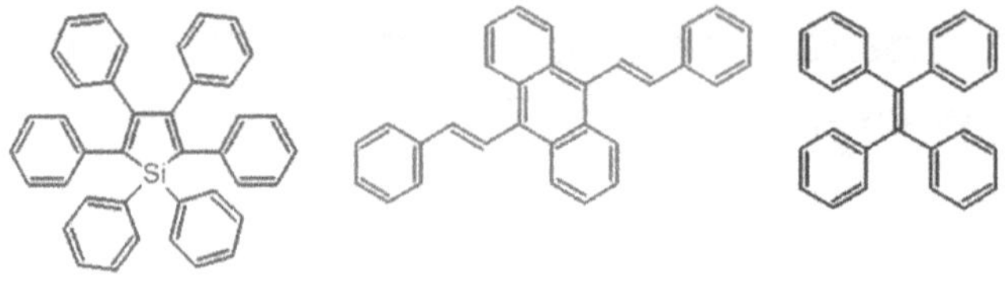

HPS DSA TPE

Scheme 1 AIE-active luminophores: Hexaphenylsilole (HPS), Distyreneanthracene (DSA) and Tetraphenylethylene (TPE), respectively.

particularly found utility in designing mechanoluminescent materials, TPE in particular attracted much attention due to their capability of self-organisation,[9] multicompound assemblies and incorporation with ACQ fluorophores. Peripheral TPE functionalisation produced a number of analogues where absorption and emission properties can be controlled in polar and non-polar solvents, depending on the type of substitution. For example, functionalisation with alkyl, aryl, oligoethylene or ACQ-active fluorophore produced non-fluorescent or weakly fluorescent compounds in solution, while giving high fluorescent quantum yield in the aggregated and solid states.[1,2,5,6] This synthetic strategy offers an important avenue to explore new AIE luminogens, which can be afforded through functionalization or decoration of TPE with ACQ fluorophore converting the whole system into applicable new AIE luminogens. Tang and co-workers have employed this strategy to functionalize typical ACQ chromophores such as anthracene (An), pyrene (Py), carbazole (Cz) and triphenylamine (TPA) with attachment of the AIE-active TPE, generating series of TPE–An, TPE–Py, TPE–Cz and TPE–TPA AIE luminogens.[10]

2 AIE-active TPE luminogen

TPE (**1**; Scheme 2) is a propeller-shaped AIE-active luminogen molecule with vast application potential. The nonplanar TPE molecule, bearing four phenyl rings around the central C=C double bond shows fluorescence in solid and aggregated states and non-emissive in dissolved state, mainly due to lack of restriction of intramolecular rotation (RIR).[6] Typically, TPE is non-luminescent in solution state because all four phenyl rings rotate freely providing a nonradiative relaxing pathway for excited electrons. However, upon aggregation and solid state, the intramolecular rotation of these phenyl rings of TPE are inhibited, preventing the effective nonradiative relaxation and resulting in strong photoluminesce as illustrated in Fig. 1. Sun *et al.*[11] investigated the dynamics of AIE of dimethoxy-TPE in water solution using quantum mechanics and molecular mechanics simulations. They found that the aggregates' size is dependent on concentration, where at low concentration the aggregates show isolated block structures whereas at high concentration, large aggregates formed with chain-type structures. The molecular environment, including solvent molecules and other solutes, determines the aggregation process. Based on the molecular dynamics simulation, small molecular aggregates start to appear in nanosecond time scale at low

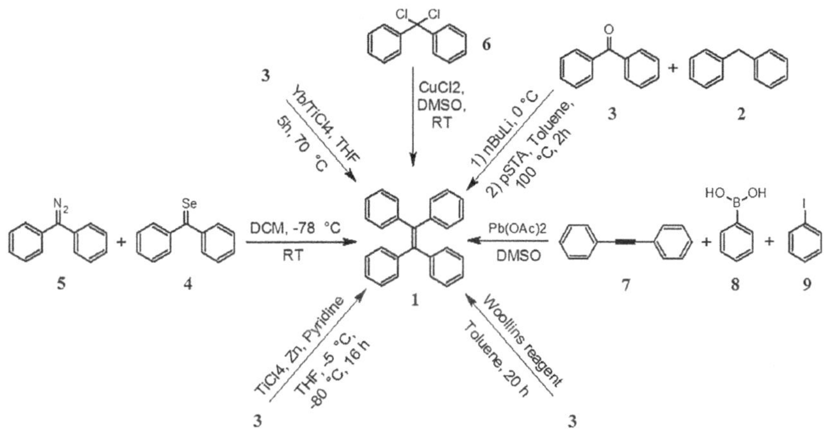

Scheme 2 Routes for the synthesis of tetraphenylethylene (TPE) luminophore.

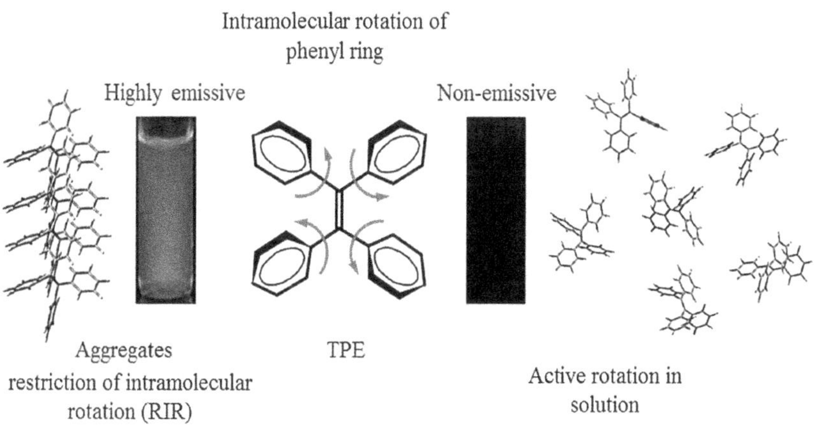

Fig. 1 AIE phenomenon of a propeller-shaped luminogen of TPE through RIR. Adapted from ref. 6 with permission from the Royal Society of Chemistry.

concentration; while the large aggregates with a chain-type structure appear at high concentration, where RIR is largely restricted due to the formation of stacked self-assembly. In the aggregated form the molecular symmetry is highly restricted in comparison to the *in vacuo* model, and the intermolecular distance is slightly shorter than the distance in the crystal structure. This study also demonstrated that the effect of electronic excitation on molecular geometry is much more significant on the isolated molecules in comparison to molecules in aggregates. The intramolecular rotation of phenyl rings is limited within 2–10° angles in comparison to the 20–55° *in vacuo*.

It is evident that substantial progress has been reported regarding functionalization of AIE-active TPE molecules. However, ACQ fluorophore functionalised with TPE luminophore and TPE derivatives has not explored for supramolecular self-assembly to date. In this chapter, we will summarize the research efforts directed to supramolecular chemistry of AIE active TPE molecules, offering a more subjective view on this fields' molecular design, self-assembly and application of AIE compound in biology and solar cells. We will conclude with potentials and prospects of TPE luminogens for future applications.

3 General syntheses and reactivity

TPE (**1**) can be prepared *via* various synthetic routes as shown in Scheme 2. The first method described by de Boissieu in 1888,[12] where they have brominated the diphenylmethane (**2**), that is subsequently heated, distilled and recrystallised from benzene to achieve **1** in reasonable yield. However, most commonly used method for preparation of TPE was reported by Wang *et al.* using Knoevenagel condensation reaction, starting with the reaction of *n*-BuLi with diphenylmethane (**2**) at 0 °C followed by the addition of benzophenone (**3**) to form tertiary alcohol which then further dehydrated using *para*-toluene sulfonic acid (pTSA) in toluene.[13] McMurry coupling is another widely used method for TPE synthesis along with its various derivatives. $TiCl_4$ and Zn powder were used in dry THF at −5 °C which mediates the reductive coupling of benzophenone (**3**).[14] This method has been used to synthesise symmetrical and asymmetrical TPE derivatives. In another report, Woollins *et al.* described heating two equivalents of ketone in toluene in the presence of Woollins' reagent to produce **1**.[15] Similarly, ytterbium (Yb) in combination with $TiCl_4$ (cat.) in THF,[16] and Heck reaction using phosphapalladacycle catalyst[17] are very effective routes for the production of TPE as well. Moreover, Okuma *et al.* described the reaction of selenobenzophenones (**4**) with diphenyldiazomethane (**5**) in DCM at −78 °C to afford good yield of TPE. They also showed that the reaction of benzophenone hydrazone with diselenium dibromide using triethylamine at room temperature gives TPE in good yield.[18] Recently, Tezuka *et al.* reported that copper catalysed self-coupling of dichlorodiphenylmethane (**6**) in DMSO at room temperature resulted in **1**.[19] Another approach to synthesise TPE was the double arylation reaction of diphenylacetylene (**7**) using phenylboronic acid (**8**), iodobenzene (**9**) and palladium acetate in DMSO.[20]

4 Self-assembly of PDI, NDI and porphyrin bearing TPE moieties

Supramolecular self-assembly of photoactive molecules has generated much interest due to possible applications in organic electronics, photovoltaics, catalysis, mechanochromic luminescence and medicinal chemistry. These self-assemblies include 1D to 3D nanostructures that generated from conventional dyes such as naphthalene diimides (NDIs), perylene diimides (PDIs) and porphyrins. Self-assemblies form *via* intermolecular interaction forces such as π–π stacking, hydrogen bonding, van der Waals, electrostatic interactions and solvophobic interactions. However, these conventional dyes suffer from drawbacks due to ACQ,[21] attractive dipole–dipole interactions and/or effective intermolecular π–π-stacking, thus inhibiting their applications as a solid or an self-induced aggregate. In contrast with NDIs, PDIs, and porphyrins, TPE and its

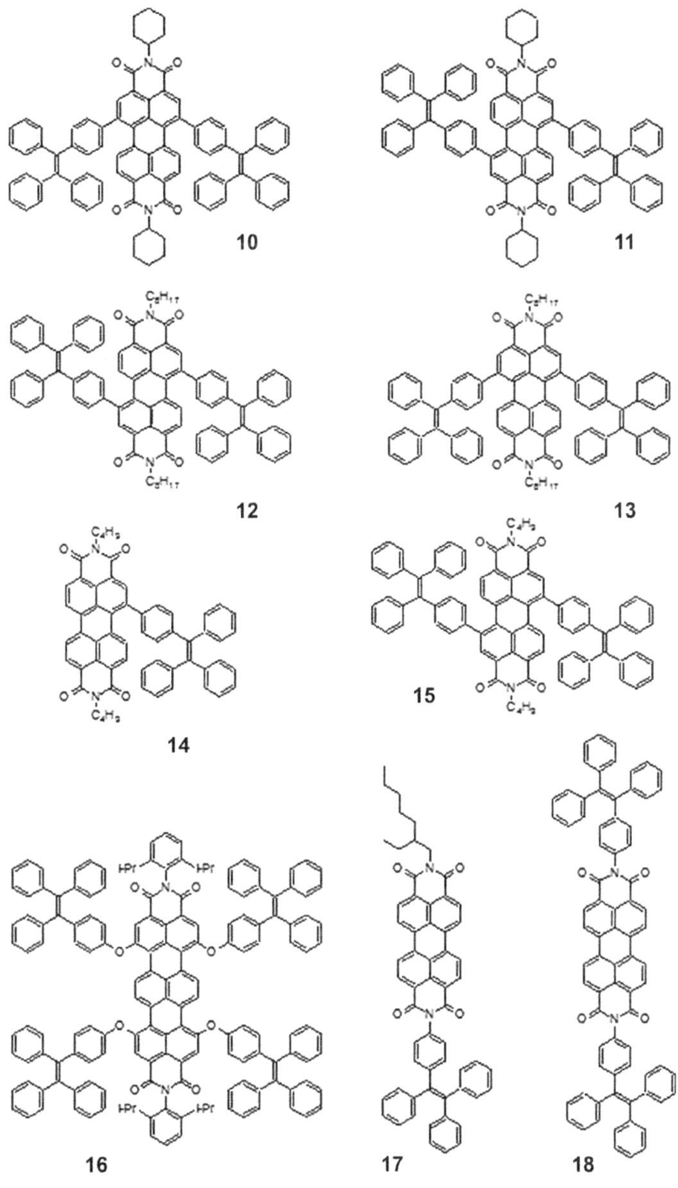

Scheme 3 A few examples illustrated the conjoint use of PDI (**10–18**) functionalised with TPE moieties.

derivatives are typical AIE-active luminophores, which undergo self-assembly in aqueous as well as organic solvents. Schemes 3 and 4 illustrate examples of porphyrin, NDI, and PDI moieties with TPE substitutions designed to utilise the properties of these units in self-assembling behaviour, and optical and photo-physical properties.

4.1 PDI-TPE

PDI, an n-type organic semiconductor, has been extensively investigated for organic light harvesting systems,[22,23] organic field transistors[24] and

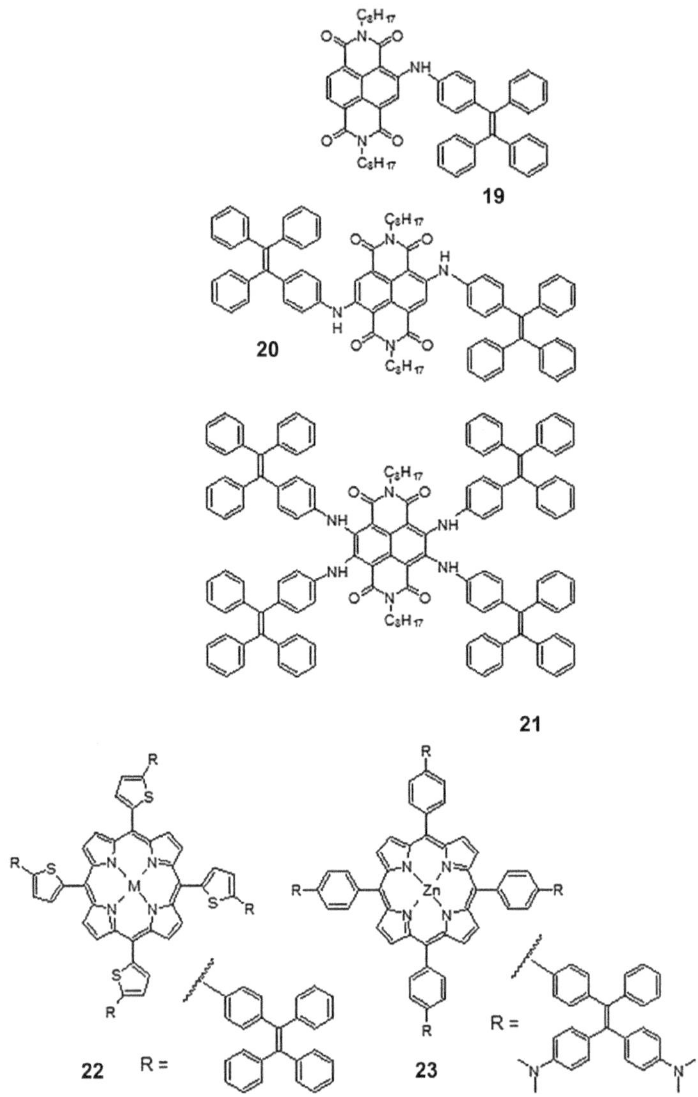

Scheme 4 Molecular structures of TPE units bearing NDI and porphyrin as electron acceptors.

biosensors[25] due to its high optical and chemical stability. PDI-based dyes have high fluorescence quantum yield in solution. However, because of strong intermolecular π–π interactions, aggregated PDIs reveal a very weak emission due to ACQ. To overcome these issues PDIs were functionalised with AIE-active TPE luminogen either through imide- or core-substitution, for example **10** and **11**,[26] **12** and **13**,[27] **14** and **15**,[28] **16**,[29] and **17** and **18**,[30] as depicted in Scheme 3.

In this regard, Zhao *et al.*[26] described the synthesis of 1,6-DTPEPDI (**10**) and 1,7-DTPEPDI (**11**) isomers by chemical modification of the PDI core with two TPE units at the bay positions. An important point to note is that the full separation of isomers of 1,6- and 1,7-substituted PDIs is

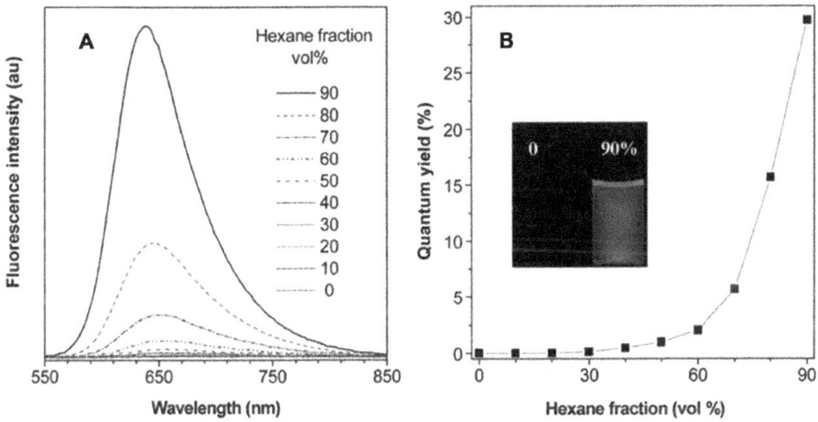

Fig. 2 (A) Fluorescence spectra of 1,7-DTPEPDI (**11**) in hexane/DCM mixtures with various hexane fractions (f_h; $\lambda_{ex} = 452$ nm). (B) Quantum yield of **11** in hexane/DCM mixtures with various f_h estimated using Rhodamine B as standard. Inset: fluorescence images ($\lambda_{ex} = 365$ nm) of **11** (10^{-5} M) in hexane/DCM mixtures with f_h values of 0 and 90%. Adapted from ref. 26 with permission from the Royal Society of Chemistry.

very challenging, however, two regioisomers **10** and **11** were easily separated by column chromatography by eluating with DCM/petroleum ether. This was explained by substitution at **10** and **11** with enhanced differences in molecular shape and symmetry. Both **10** and **11** derivatives were non-luminescent in good organic solvents such as THF, $CHCl_3$, and DCM, however, 530-fold higher fluorescence was observed when bad solvents such as hexane, methanol and water were added to good solvents (Fig. 2). Interestingly, the $\Phi_{F,solid} = 6.3\%$ of **11** in solid film is 90 times higher than that of dilute solution ($\Phi_{F,solute} = 0.07\%$). This clearly demonstrates transition of PDIs from ACQ to AIE behaviour. The group also investigated supramolecular assembly of **11** in various non-polar/polar solvent mixtures, obtaining a range of ordered and well-organized 1D nanostructures. For example, assembled **11** nanofibers can be easily formed in mixtures of hexane/CH_2Cl_2, methanol/CH_2Cl_2 (Fig. 3A), water/THF (Fig. 3B), methanol/THF (Fig. 3C) and methanol/dioxane (Fig. 3D). Wire- and rod-like structures were assembled by slow evaporation of solvent mixtures (Fig. 3E and F). The electronic structure of **11** remains unchanged in comparison to the PDI core, maintaining its typical n-type organic semiconductor behaviour, and only a red-shift in the fluorescence spectrum can be observed due to TPE decoration to the PDI core.

In another study, Aldred *et al.*[27] found that the fluorescence quantum efficiency of PDI-TPE (**12, 13**) not only depends on aggregation but also on the morphology of such aggregates. In the amorphous state, PDI-TPE displays an emission at 744 nm, while gives a blue shift in emission to 665 nm in the crystalline state. This phenomenon leads to a piezo-fluorochromic behaviour in which emission property can be tuned by grinding and thermal annealing. The crystalline microstructures allow fabricating microwire field-effect transistor (MFET) with the effective mobility (m_e) and capacitance (C_i) estimated to be 3.11×10^{-5} cm^2 V^{-1} s^{-1} and 1.51×10^{-15} F, respectively.

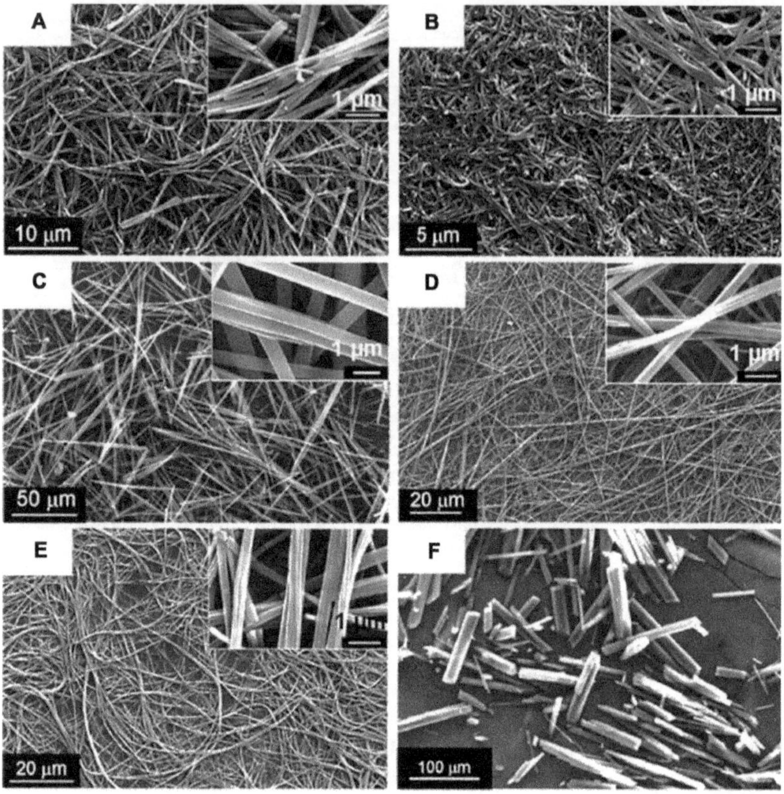

Fig. 3 SEM images of the morphologies of aggregates formed by 1,7-DTPEPBI (**11**) molecules at 10^{-5} M concentration; (A) methanol/DCM mixture ($f_m = 70\%$), (B) water/THF mixture ($f_w = 60\%$), (C) methanol/THF mixture ($f_m = 90\%$), (D) methanol/dioxane mixture ($f_m = 80\%$), (E) fibrils formed by slow solvent evaporation from water/THF mixture ($f_w = 40\%$) solution, and (F) rod- and prism-like microstructures formed by slow solvent evaporation of **11** from THF solution. Adapted from ref. 26 with permission from the Royal Society of Chemistry.

Zhao *et al.*[28] findings reveal that the number of TPE substituents on PDI cores have a significant effect on AIE behaviour. Unlike pristine PDI (non-substituted) which shows a strong ACQ effect, mono- and di-TPE substituted PDI derivatives (**14, 15**) exhibited AIE behaviour. In dilute solutions, mono- and di-TPE substituted PDIs displayed a low fluorescence efficiency with a quantum yield (Φ_{solut}) of 2.2% and 0.07%, respectively. However, the quantum efficiency reasonably increased to 9.0% for mono-TPE-PDI and 18.9% for di-TPE-PDI in aggregate forms.

In a similar fashion, Xie *et al.*[29] synthesised two novel terylene diimide (TDI) dyads bearing four TPE moieties, linked to the core with (**16**) or without oxygen, and exhibiting strong AIE effect with very strong emission at near-infrared (NIR) region around 800 nm. These novel AIE-active NIR emitters will be useful for future applications in NIR organic light emitting diodes (OLED) and in *in vivo* bio-imaging. Recently, Wang[30] *et al.* synthesized TPE-N-PDI (**17**) and DTPE-N-PDI (**18**) by attaching one and two TPE moieties *via* an imide-linkage (Scheme 3). The solutions of both **17** and **18** are weakly emissive, while the solid films and aggregates formed by **17** and **18** exhibit distinct AIE activity, while PDIs with alkyl-, alkyloxy-, and phenyl substitutions with imide linkage show ACQ effect. Such phenomenon is due to the photo-induced electron transfer between TPE and PDI subunit. The self-assembly of both **17** and **18** can be manipulated using solvent mixtures. In DCM/hexane mixtures, both **17** and **18** assembled to form X-aggregates, while H-type assembly was observed with **17** in THF/water mixture due to thinner alkyl chain. These studies show the possibilities in tailoring and engineering the AIE and ACQ active molecules and using the properties these moieties have to offer to balance and alter classical luminogens and create highly efficient luminescent solid materials.

4.2 NDI-TPE

NDI is a smaller analogue of PDI and TDIs, which has been extensively investigated as active component for many promising applications including, but not limited to, organic light harvesting systems,[31] medicinal[32] and transistors.[33] Although similar to PDI, NDI displays a near-unity fluorescence quantum yield in dilute solutions, however, in aggregates or solid state it emits weakly which is referred to ACQ behaviours. Such ACQ effect is ascribed to intermolecular π–π stacking interaction. In order to enhance fluorescence performance in solid and aggregated states, *i.e.* convert NDI-based derivatives from ACQ to AIE active materials, mono-, di- and tetra-TPE substituted NDIs were designed where AIE-activity of the TPE luminophore moieties is used to modify the NDI electronic transitions in aggregated state.

We reported the synthesis of two core-substituted NDI derivatives bearing one and two amino-TPE at the core, namely, 2-TPEcNDI (**19**) and 2,6-DTPEcNDI (**20**) as shown in Scheme 4.[34] Fluorescence quantum efficiency of cast film of **19** is 6.9%, which is higher than its solution in $CHCl_3$ ($\Phi_{solut} = 1.9\%$) and THF ($\Phi_{solut} = 0.9\%$). The diTPE substituted NDI **20** exhibited weak emissions with $\Phi_{solut} = 0.09\%$ in a dilute solution of $CHCl_3$, however, a strong emission in the near-IR region with a $\Phi_{solut} = 18.3\%$ in the solid state was observed for **20**. The effect of solvents on **19** and **20** aggregation and fluorescence performances was evaluated using Φ_{solut} in hexane/$CHCl_3$, methanol/$CHCl_3$, and water/THF solvent mixtures, which was 12.8%, 9.7% and 7.1%, respectively, for **19**, while higher Φ_{solut} of 21.9%, 15.3%, and 19.1% was observed for **20**. Furthermore, scanning electron microscopy (SEM) was used to study the effect of solvent on the self-assembly of **20** as shown in Fig. 4A–C, where various well-ordered microstructures such as fibres, needle-like and nano-belts formed *via* solvophobic control. Most recently, we evaluated the effect of number of TPE substitutions on NDI core for compounds' optical properties. Core-substituted NDIs with four amino-TPE luminophores TTPEcNDI (**21**) (Scheme 4)[35] showed a high AIE effect with $\Phi_f = 0.07\%$ in pure $CHCl_3$ solution and $\Phi_f = 21.1\%$ in solid state. The AIE effect was further studied in different solvent mixtures of hexane/$CHCl_3$, methanol/$CHCl_3$, and water/THF mixtures, which obtained the quantum yield of 29.9%, 24.7% and 16.8%, respectively. It

Fig. 4 SEM images of self-assembly formed by 2,6-DTPEcNDI **20** in (A) chloroform/hexane ($f_h = 90\%$) produces hollow spheres, (B) chloroform/methanol ($f_m = 70\%$) crystallites fibriles and (C) THF/water ($f_w = 40\%$) formed leaf-like aggregates. (D) TTPEcNDI **21** molecular self-assemblies in $CHCl_3$/methanol mixture with $f_m = 70\%$, (E) in $CHCl_3$/hexane ($f_h = 90\%$) and (F) in THF/water mixture with a water fraction ($f_w = 40\%$) at concentration of 10^{-5} M. A–C are adapted from ref. 34 with permission from the Royal Society of Chemistry. And D–F are adapted from ref. 35 with permission from the Royal Society of Chemistry.

is interesting that **21** self-assembled into different morphologies in various good/poor solvent systems *i.e.* hollow spheres in $CHCl_3$/hexane (Fig. 4D), crystallites and fibrils in $CHCl_3$/methanol (Fig. 4E), and leaf-like microstructure in THF/water (Fig. 4F).

4.3 Porphyrin-TPE

Porphyrins, which have a similar structure to chlorophylls – a compound that can convert light into chemical energy – are of interest in biology,[36] solar cell,[37] tumor sensing,[38] medical imaging[39] and optoelectronic devices.[40] However, porphyrins are ACQ active in aggregates, and TPE moieties can be used to decorate porphyrin rings to improve optical and photophysical properties, as well as fluorescence performance in the solid state. In 2015, we synthesized π-conjugated porphyrin derivatives bearing TPE units **22** (TPE-Por) as shown in Scheme 4.[41] The resulted material displayed an emission band at near-infrared region of 700 nm with high fluorescence efficiency of 15% when excited at 440 nm. It was shown that the self-assembly behaviour of **22** in polar and non-polar solvent mixtures is driven by J- and H-aggregates *via* π–π stacking interactions, and solvophobic interactions further play a significant role in defining the final morphology of aggregates as shown in Fig. 5.

In an another report, Guo *et al.*[42] have demonstrated the AIE-effect of TPE-based porphyrin derivative by investigating fluorescence performances of TPE-Por (**23**) in various THF/water mixture ratios. At low water ratios ($f_w \leq 30\%$), the emission of **23** is significantly reduced due to increase solvent polarity and intermolecular charge transfer. Further increasing of water fraction in the solvent mixture ($f_w \geq 40\%$) drove the self-assembly process and aggregate formation, resulting in a new peak

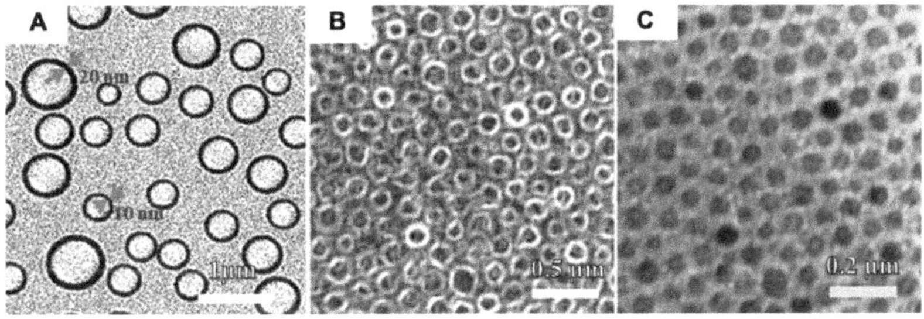

Fig. 5 TEM images of **13** (10^{-4} M) from an equimolar mixture of (A) $CHCl_3$/MeOH, (B) CH_3CN/MeOH, and (C) CH_3CN/cyclohexane, respectively. Adapted with permission from ref. 41. Copyright 2015 American Chemical Society.

at 725 nm, with increasing fluorescence efficiency transforming the porphyrin moiety to an AIE active electronic structure.

5 TPE moieties bearing macromolecules

Macromolecules can readily self-assemble into a variety of ordered and well-defined geometries and form morphologies in the size range of nm to μm. Along with outstanding AIE-effect of TPE and its derivatives, the self-assembly of such molecules to form supramolecular well defined micro and nanostructures increases their application potential. For example, TPE-based macrocyclic compounds have attracted extensive interest due to their excellent properties in molecular recognition.

In 2013, Song *et al.*[43] synthesized a TPE-based macrocycle by joining two TPE moieties with triethylene glycol chains (**24**) and investigated its self-assembly for the first time (Scheme 5). AIE behaviour was confirmed when no emission was observed in the dilute THF solution while strong fluorescence was emitted in solid state with a quantum yield of 9.1, 10 and 15% in THF/H_2O mixture of 70, 90 and 95% water, respectively. It was also noted that the morphology of self-assembled microstructures changed as a function of water fraction. At 70% water fraction, nanorods were mainly obtained (Fig. 6A). Further adding water to 90%, TPE macrocycle assembled into solid nanospheres (Fig. 6B). Interestingly, in 95% water fraction, the morphology transformed to hollow microsphere structure with a diameter of 1–8 μm with the shell width of 0.1–1 μm (Fig. 6C).

The same group further reported a new amide macrocycle (**25**) based on the TPE luminophore.[44] The resultant macrocyclic **25** displayed a typical AIE effect where no light emission was observed in dilute DMF or THF solutions, whereas high blue fluorescence emission was observed in aggregate state in a mixture of H_2O/DMF or H_2O/THF. The TPE-based amide macrocycle also demonstrated a facial assembly into a well-organized microstructure with various morphologies in H_2O/THF mixture with varying water fractions. In 70–80% water in THF mixture, **25** assembled into microtubes with diameter ranging 0.3–3.0 μm, wall thickness of 0.1–0.5 μm and length of 0.3–4 μm formed by the self-rolling of nanoribbon precursors (inset in Fig. 6D). However, with water fraction

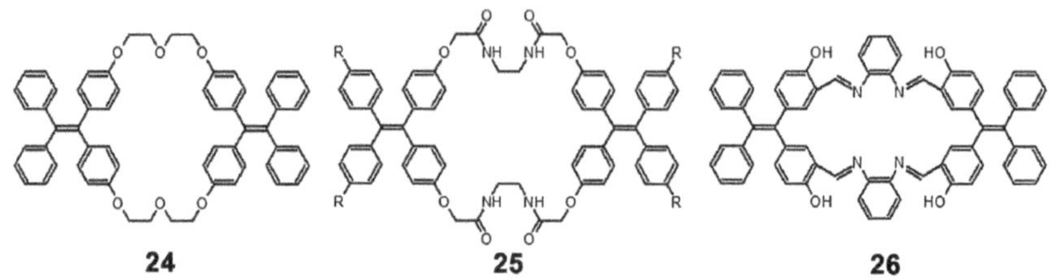

Scheme 5 Chemical structures of TPE-based macrocyclic compounds.

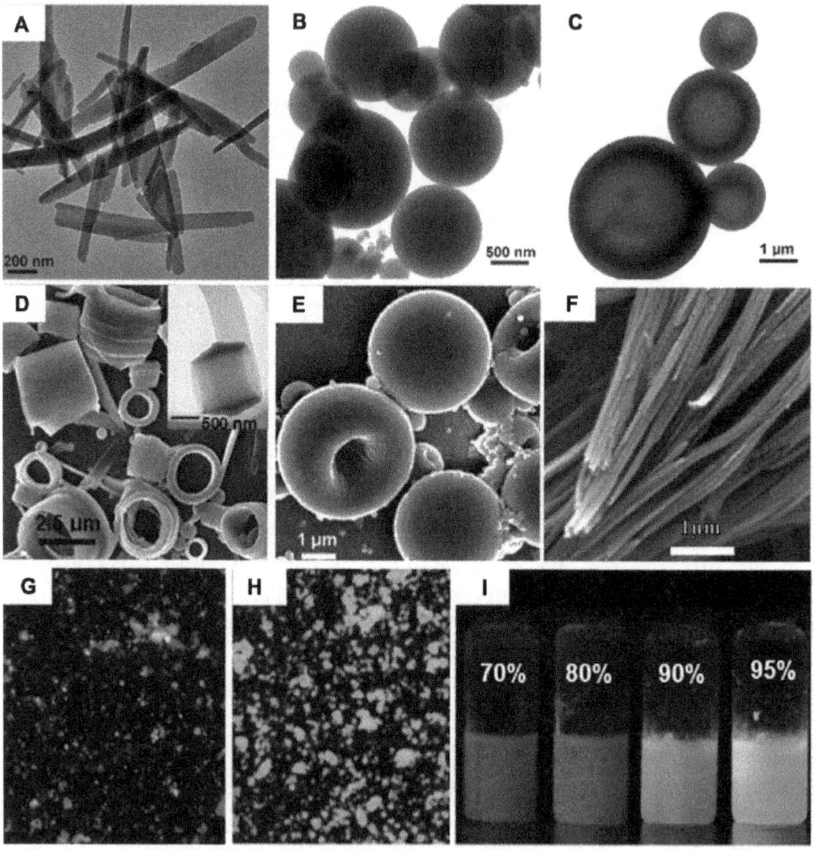

Fig. 6 Electron microscopy images of a suspension of **24** in water and THF with different water percentages: (A) in 70% water, (B) in 90% water, (C) in 95% water from 2.5×10^{-3} M solution. FE-SEM images of the suspension of 25 in (D) 70%, and (E) 95% H_2O–THF with water fractions, $[25] = 1.0 \times 10^{-3}$ M. (F) FE-SEM image of **26** suspension from 0.1 mM solution in 90% H_2O–THF. Inverted fluorescence microscope images of dry powder of **25** obtained by centrifugation of suspensions in H_2O/THF with water of 70% (G), 95% (H). (I) Photograph of a suspension of **25** in H_2O–THF with increasing water fraction under irradiation of a UV lamp at 365 nm. A–C are adapted with permission from ref. 43. Copyright 2013 American Chemical Society and D–F are adapted from ref. 44 with permission from the Royal Society of Chemistry.

higher than 90% hollow microspheres were produced (Fig. 6E). The microtubes produced by self-rolling of big nanoribbons emitted blue light while the hollow microspheres gave blue-green fluorescence due to conformation difference of the macrocycle **25** in the supramolecular structure (Fig. 6G–I).

In the same year Feng *et al.*[45] synthesised a Schiff-base TPE macrocycle (**26**) using the condensation reaction of TPE dialdehyde and 1,2-benzenediamine which showed a typical AIE-effect with strong red-light fluorescence in THF/H_2O mixture, while no emission was observed in pristine THF solution. The fluorescence intensity of the suspension increased with increasing water fraction where **26** pushed to self-assemble due to solvophobic effects. In the mixture of THF/H_2O with the volume ratio of 1 : 9, macrocycle **26** self-assembled into nanofiber structure with the diameter of 100–200 nm (Fig. 6F). This form of aggregation showed a large Stoke shift up to 260 nm in emission, primarily due to AIE effect. The group used **26** macrocycles for the detection of Cu(II) in real water samples as well as in pork juice-containing water by naked eye with a limit of detection of 10 μM.

Apart from the AIE macrocyclic compounds, many macromolecules are proposed which not only exhibited outstanding AIE behaviour, but also self-assembled into ordered and well-defined nano/microstructures. A series of such compounds is shown in Scheme 6. In 2015, we designed a rigid supramolecular star-shaped luminogen **27** (HTCA) by anchoring six TPE units to the core of cyclohexanehexone ring *via N*-atoms showing typical AIE effect.[46] The enhanced emission of **27** luminogen in aggregates was determined to be 218 times higher than that of its dilute solution. The self-assembled morphologies of the HTCA aggregates change with water fraction in the water/THF mixture. Typically, cross-like

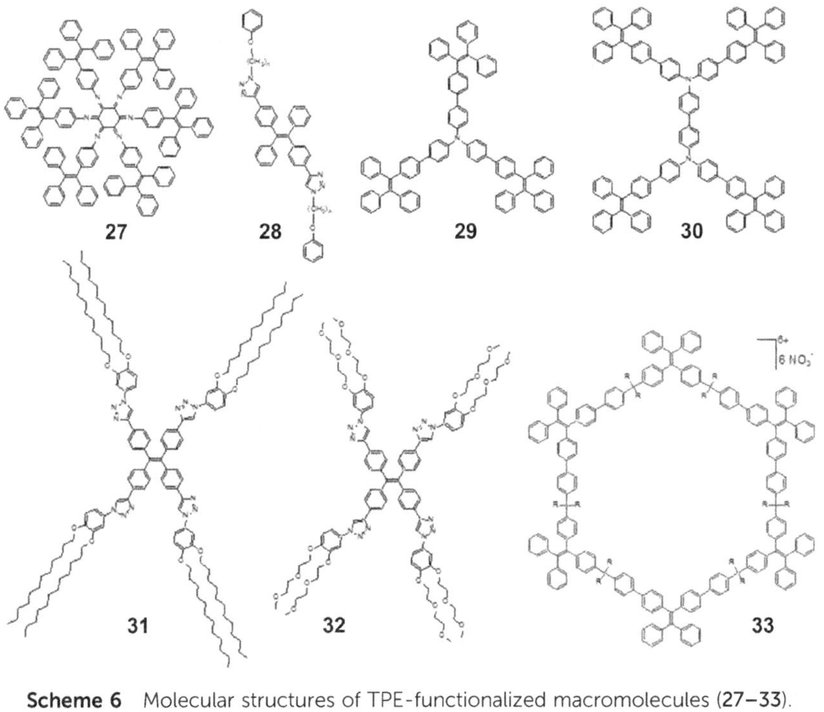

Scheme 6 Molecular structures of TPE-functionalized macromolecules (**27–33**).

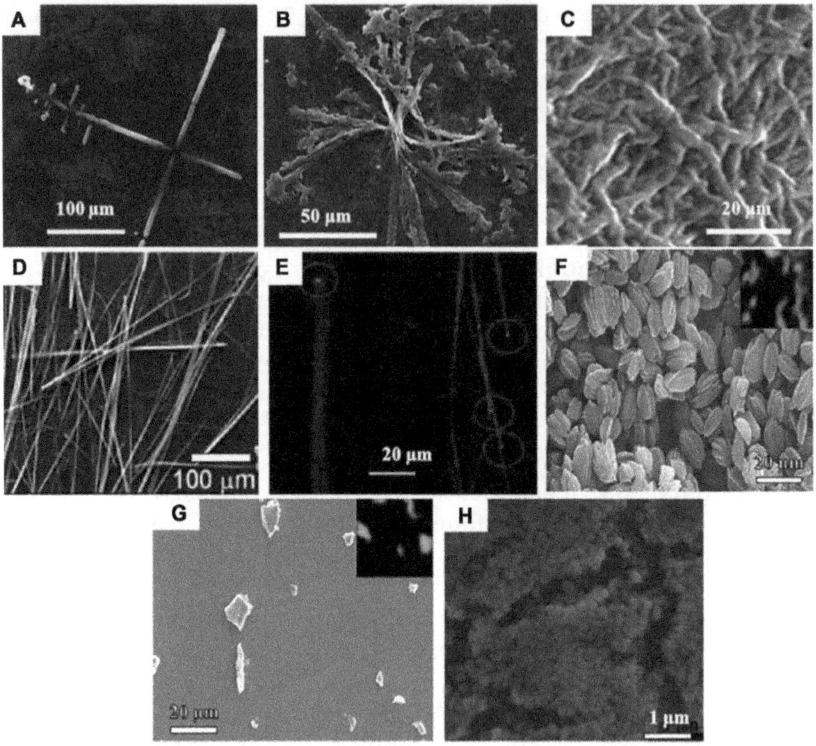

Fig. 7 SEM micrographs of HTCA (10^{-5} M) from water/THF at fractions of $f_w = 50\%$ (A), $f_w = 70\%$ (B) and $f_w = 90\%$ (C), respectively. (D) SEM image of 33 in CH_2Cl_2/hexane mixed solutions with 90% hexane content. (E) SEM of the self-organized structures of 28 formed in the THF/water mixtures of $f_w = 60\%$ after standing for 1 week, and (F) Fluorescent image of the microfibers taken under a fluorescence microscope with a 337 nm excitation. SEM images of (G) the aggregates of 32 at $f_{(hexane)} = 80\%$, and (H) the aggregates of 32 $f_{(hexane)} = 90\%$. The inset images are the fluorescence images of 32 under UV illumination at 365 nm. A–C are adapted from ref. 46 with permission from the Royal Society of Chemistry, D & E are adapted with permission from ref. 47. Copyright 2012 American Chemical Society, F & G are adapted from ref. 49 with permission from the Royal Society of Chemistry and H is adapted with permission from ref. 50. Copyright 2016 American Chemical Society.

rod shapes with a diameter of 5–10 μm were formed in THF/H_2O with water fraction of 50% (Fig. 7A). However, when the water fraction was increased to 70%, HTCA was assembled into crossed-fibril microstructures (Fig. 7B). Interestingly, intertwisted helical aggregates with the average width between 0.5 and 2.0 μm and the length of several micrometers were obtained with high water fraction of 90% (Fig. 7C).

A remarkable AIE active supramolecular structure with fluorescence quantum yield of ~100% of TPE derivatives was reported by Wang and co-workers.[47] In their study, a mixture of *E* and *Z* isomers, triazole-functionalized-TPE derivative named 1,2-bis{4-[1-(6-phenoxyhexyl)-4-(1,2,3-triazol)yl]phenyl}-1,2-diphenylethene (BPHTATPE) (**28**) was synthesized *via* click chemistry using Cu as a catalyst. The obtained *E* isomer of BPHTATPE assembled into ordered and well-organized 1D nanostructures of microfibers (Fig. 7D) and nanorods in THF/water and hexane/$CHCl_3$, respectively. The microfibers of **28** showed waveguide effect, usually observed in optically anisotropic systems, which we can see in the *E* isomer of **28** forming microcrystals to self-organize. The *Z* isomer, on the other hand, can hardly form any microcrystalline structures, due to its irregular conformation. The optical waveguiding effect offers the possibility of constructing advanced photonic devices using these microfibers as the building blocks (Fig. 7E).

In 2010, supramolecular structures formed from 3TPETPA **29** and 4TPETPA **30** compounds were fabricated by bearing three and four TPE moieties to triphenylamine (TPA) and $N^4,N^4,N^{4'},N^{4'}$-tetraphenylbiphenyl-4,4′-diamine (DTPA), respectively.[48] The obtained luminogen exhibited an excellent transformation from ACQ to AIE behaviour. The $\Phi_{F,solut}$ of 3TPETPA was 0.42% in the dilute solution and $\Phi_{F,solid} = 91.6\%$ in the solid state. A similar AIE effect was also observed for **30** with $\Phi_{F,solut} = 0.55\%$ in solution and $\Phi_{F,solid} = 100\%$ in the solid state.

Two luminogens designed with hydrophobic dodecyl **31** or hydrophilic di(ethylene oxide) **32** peripheral chains and an identical extended TPE aromatic core were later published in 2015 by the Han group.[49] In dilute chloroform solution, the luminogens were almost similar in photoluminescent (PL) spectra with weak fluorescence intensities. However, with the addition of methanol (70–80%) or hexane (80%) to $CHCl_3$ solution, a bright bluish colour was observed with high fluorescence intensity in the PL spectra. The aggregate morphology of luminogens in mixed methanol/$CHCl_3$ solution indicated that aggregates were crystalline with a needle-like structure. However, aggregates in hexane/$CHCl_3$ mixture revealed rice seed-like microstructures (Fig. 7F), with further increase of hexane fraction up to 90% no obvious morphology was observed (Fig. 7G).

Yan and co-workers describe the synthesis of TPE-based di-Pt(II) acceptors **33**, which used to construct 2D hexagonal metallacycles and 3D drumlike metallacages with three different counteranions *via* coordination-driven self-assembly.[50] The structure of these derivatives composed of alternating donor (TPE) and acceptor (di-Pt(II)) units with AIE behaviour. These metallacycle luminogen were weakly emissive in dilute solution and emitted strongly in solid form. In addition, strong AIE effect was observed using a higher fraction of hexane in CH_2Cl_2 (90%, v/v) with increasing solvophobic effects. At this solvent ratio, metallcycles assembled into uniform and well-defined nanospheres with average diameters ranging 150–250 nm (Fig. 7H). The morphologies of aggregated nanostructures remained unchanged with further increase of hexane fraction due to the rigid nature of the molecular geometry.

6 Chiral assembly

The design and synthesis of TPE-based organic molecules, which can both assemble into helical micro/nanostructures with selective handedness, and give AIE behaviour is of interest due to potential applications in sensors, liquid crystals, optical devices[51–53] and as probe to study mainly chiral life processes. Supramolecular helicity can be simply induced by assembly of molecules containing chiral centres such as peptides, saccharides *e.g.* glucose, π-conjugated oligomers and bola-amphiphiles through intermolecular π–π stacking, solvophobic and other interactions. Chiral amino acids with strong supramolecular self-assembling capabilities, which can lead to well-defined assembled structures, are ideal components to induce chirality. In 2014, Li *et al.*[54] reported a val-TPE **34** supramolecule by hybridizing valine (as a chiral block) with AIE-active TPE moiety using copper-catalyzed azide–alkyne "click" reaction (Scheme 7). The optical properties and self-assembly were determined by both the AIE of TPE and amino acid moieties. In THF solution, val-TPE exhibited no signal of emission as well as very weak circular dichrosim (CD) signal. However, the fluorescence intensity increased significantly in THF/water mixture with strong CD signal. Fig. 8C shows a bright blue luminescence image of **34** in its solid state. This phenomenon confirms that val-TPE is a highly AIE active compound and at the same time when molecules self-assemble, assumes a fixed position where chiral superstructure can be detected as is evidenced by a CD signal in fluorescence wavelength range. The self-assembly of val-TPE in 1,2-dicloroethane/hexane mixture was examined by SEM and TEM as shown in Fig. 8A and B. It is clear that val-TPE assembled into a left-handed twisted

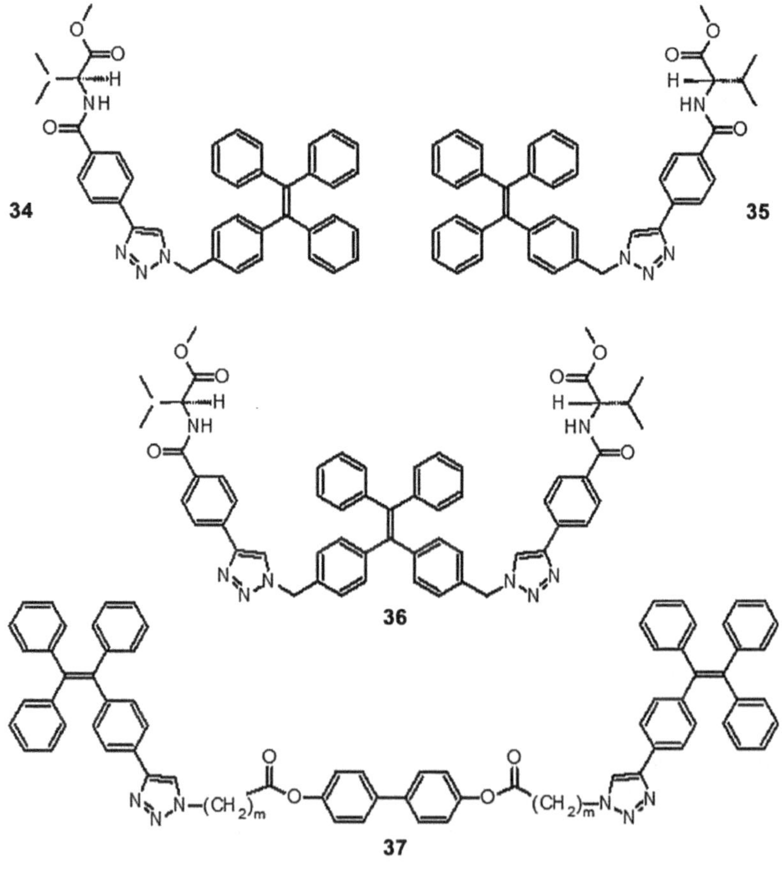

Scheme 7 Molecular structures of TPE-based chiral center-containing molecules.

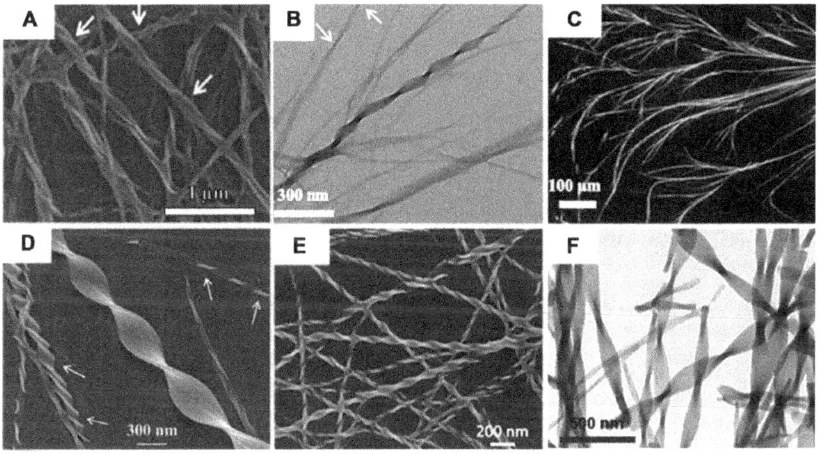

Fig. 8 SEM (A) and TEM (B) of the aggregates of Val-TPE **34** formed upon the evaporation of 1×10^{-4} M solution in DCE–hexane (1/9, v/v). (C) Fluorescent image of **34** upon the evaporation of DMF solution on quartz substrates in the micropatterned channel under UV excitation. (D) SEM images of the aggregates of **35** formed in DCE–hexane (1/9, v/v) mixture and 10^{-4} M concentration. (E) SEM images of **36** formed by solvent evaporation of its solution of 1×10^{-4} M in DCE. (F) TEM images for the aggregates of **37** ($m = 5$) in THF–water mixture (40/60) formed upon standing for 6 h. A–C are adapted from ref. 54 with permission from the Royal Society of Chemistry, D is adapted from ref. 55 with permission from the Royal Society of Chemistry, E is adapted with permission from Macmillan Publishers Ltd: Scientific Reports (ref. 56). Copyright 2016 and F is adapted from ref. 57 with permission from the Royal Society of Chemistry.

nanofibers with an average width between 35–55 nm and a helical pitch of 190 nm. The formed helical structure is consistent with the positive signal from CD spectrum.

Similarly, Li and co-workers synthesised L-leucine methyl ester coupled with TPE moiety to form a chiral AIE active TPE-Leu **35** derivative.[55] In the dilute THF solution, **35** is non-emissive with silent CD behaviour. However, it had intense blue emission and strong CD signals in an aggregate state, which clearly demonstrated that **35** is typical AIE-active and produces chiral self-assembly. In 1,2-dichloroethane/hexane mixture, **35** assembled into left-handed helical nanofibers and nanoribbons with a broad width distribution of 15–350 nm as well as a wide range of helical pitch up to 920 nm as shown in Fig. 8D. These nanofibers and nanoribbons exhibited circularly polarized luminescent (CPL) properties with large CPL-emission dissymmetry factor of 0.02–0.07.

In another example, Li *et al.* designed and synthesized chiral TPE derivative with two valine-containing attachments (TPE-DVAL) (**36**) through the azide–alkyne click reaction.[56] The two amino acid pedants were introduced to TPE scaffold to form a new chiral TPE derivative. TPE-DVAL **36** is non-luminescent and CD silent in dilute DCM and DCE solutions, but displays a typical AIE behavior and good CD signal, where it shows strong fluorescent and Cotton effect in DCM/hexane mixture and solid film formed on a quartz substrate, respectively. The self-assembling of **36** was investigated on microstructures deposited by solvent evaporation from pure DCE solution as shown in Fig. 8E. TPE-DVAL **36** self-assembly produced uniformly right-handed helical twisted nanofibers with an average width of 30–70 nm, and helical pitch of 100–200 nm and the average length up to several micrometres. These results suggested that the self-assembling behaviour and optical properties can be tuned by changing the number of amino acid attachments.

Another twisted helical nanostructure can also be formed by self-assembly of molecules in the absence of chiral centers. Yuan *et al.*[57] designed and synthesized a TPE-based triazole compound **37** *via* Cu(I)-catalysed azide–alkyne "click" reaction. This **37** luminogen showed strong AIE in aggregate suspension in THF/H_2O mixture with an increase in emission of 5944 times in comparison to its dilute solution in pure THF. In THF/H_2O solution with $f_w = 60\%$, **37** assembled into well-defined right-handed helical nanofibers with an average width of ~100 nm and up to several micrometers in length (Fig. 8F). The comparative study of flexible spacers indicates its crucial importance in the formation of such helical structures. The self-assembly and gelation of the **37** ($m = 10$) occurs at much lower concentration in comparison to **37** ($m = 5$), an evidence that longer spacer length is favoured for aggregation and self-assembly process. Removing the alkyl chain from spacer resulted in much more rigid molecular structure and self-assembled to form crystalline particles or nanosheets.

AIE-active TPE luminogens can be designed to give fluorescent light-up bioprobes where fluorescent probes are activated based on interactions between probes and active agents in the biological environments. Peptides are the most used biological building blocks to modify TPE moieties to form AIE bioprobes such as TPE probes with peptide-linkage moieties facially bind to analyte upon presences in the environment. For instance, the salt-responsive Q11 peptides capped with TPE unit **38, 39** (Scheme 8) were successfully synthesized and characterized.[58] The incorporation of Q11 section in the peptide made conjugate aggregation, controllable gel formation, and provided mild gelation conditions which would be critical for tissue engineering and biosensing. Zhang *et al.* found that the AIE active peptide conjugates are turned on only in the presence of salt where the monodisperse hydrogelators self-assemble into a hydrogel network with bright emission rather than temperature, pH, or solvents. The 3D networks of nanofibers were observed in hydrogel formation upon addition of salts as shown in Fig. 9A. A range of TPE-capped dipeptides (TPE-GG) (**40**) were also prepared using a solid phase peptide synthesis varying the side moieties on the dipeptide section.[59] These compounds self-assembled to form gels with nanosheets with an average width of approximately 300 nm. Hydrogel formation was observed by TEM and AFM imaging as shown in Fig. 9G and **H**. The small size of peptide resulted in more rigid order molecular assembly in the hydrogel.

Han *et al.*[60] reported AIE light-up bioprobes which were constructed from a TPE moiety and GFFY self-assembling peptide, and linked to other peptides that can be the recognition elements of the bioprobe (**41, 43**). Analogue structures were also synthesised without the GFFY moiety as the control probes (**42, 44**) as shown in Scheme 8. Thanks to the hydrophilicity of peptide sequences, peptides–TPE conjugates were non-luminescent in aqueous solution. However, in the presence of analytes

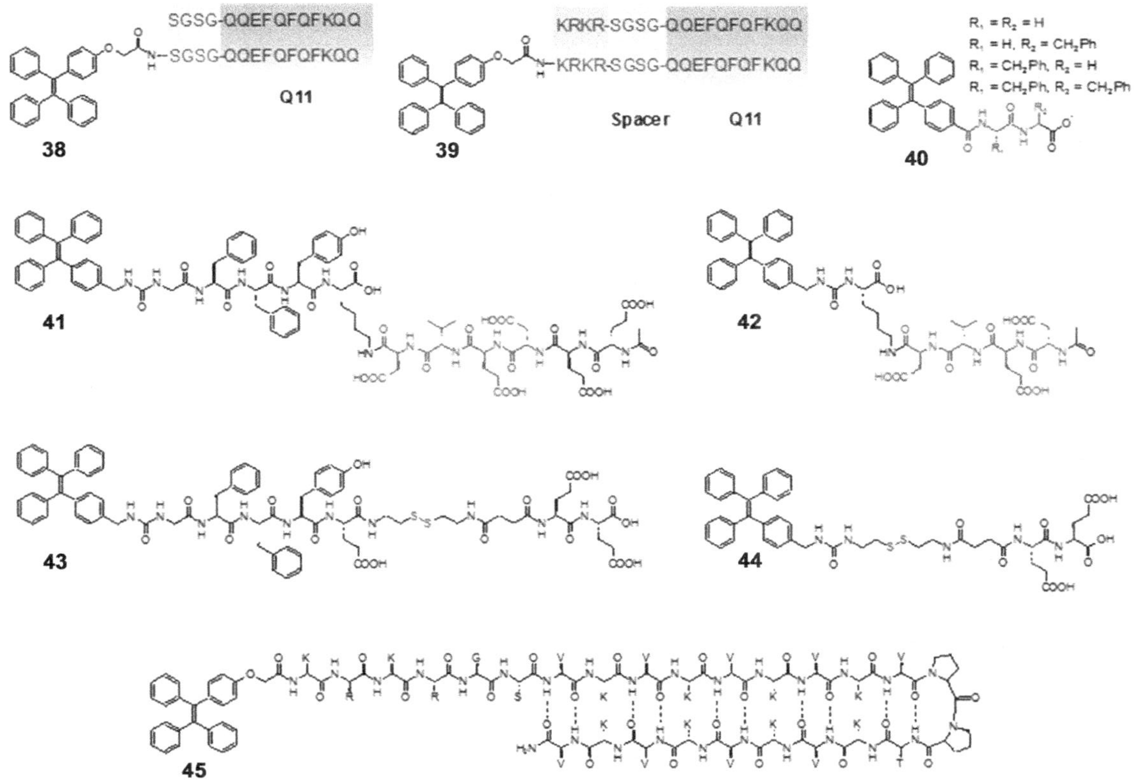

Scheme 8 Chemical structures of peptides and TPE-peptide used to form molecular hydrogel (38 and 39). The chemical structures of TPE-GG, TPE-GF, TPE-FG, and TPE-FF (40), TPE-GFFYK(DVEDEE-Ac) (41), TPE-K(DVED-Ac) (42), TPE-GFFYE-SS-EE (43), TPE-SS-EE (44), and TPE-MAX (45).

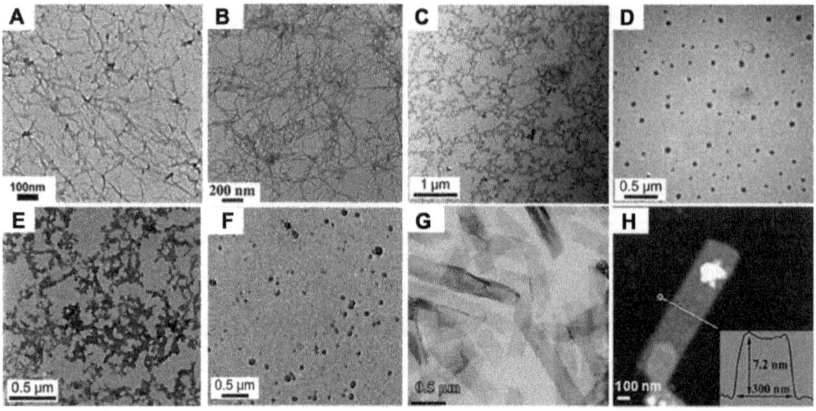

Fig. 9 (A) TEM images of TPE-Q19 (**39**) gel at 0.5 wt% concentration. (B) TEM image of TPE-MAX (**45**) gel. TEM images of (C) TPE-GFFYK-(DVEDEE-Ac) (**41**) and (D) TPE-K(DVED-Ac) (**42**) aggregates after treatment with caspase-3. TEM images of (E) TPEGFFYE-SS-EE (**43**) and (F) TPE-SS-EE (**44**) after treatment with GSH. (G) Negative stained TEM image and (H) AFM measurement of hydrogels of **40** ($R_1 = R_2 = H$) at the concentration of 3 wt%. A–C are adapted with permission from ref. 58. Copyright 2013 American Chemical Society, G & H are adapted from ref. 59 with permission from the Royal Society of Chemistry, C–F are adapted with permission from ref. 60. Copyright 2016 American Chemical Society and B is adapted from ref. 61 with permission from the Royal Society of Chemistry.

(here, caspase-3), which were ascribed to cleave the hydrophilic moieties and accordingly induced aggregation in the TPE-GFFY segment, resulting in strong fluorescence. The morphologies of aggregates were either filamentous network nanostructures or nanoparticles depended on types of peptide attachments (Fig. 9C–F).

In 2015, Zhang *et al.*[61] prepared a peptide-capped TPE luminogen (TPE-MAX) (**45**) with highly sensitive response to pH through a solid phase peptide synthesis. In dilute aqueous solution with pH 6.0, TPE-MAX was non-emissive. When pH increased from 8.0 to 10.0, the intensity of fluorescence significantly increased due to self-assembly, evidence of its AIE activity. Upon hydrogelation, a network architecture of nanofibers with a diameter of approximately 10 nm is formed (Fig. 9B).

So far, we have seen TPE core-based compounds that demonstrate the formation of twisted fibres from either homochiral molecules or amino acid/peptide coupled TPE derivatives, where the molecular chirality is transferred to the handedness of self-assembled helical structure. However, the induction and controlled chirality (non-racemic) of the supramolecular assembly from achiral AIE-active TPE molecules are quite challenging. In this regards, we reported a chiral supramolecular twisted assembly based on hierarchically organized supramolecular micro-architecture **46** from an achiral TPE derivative using solvophobic control (Fig. 10).[62] The design of this AIE-active molecule includes a TPE-core functionalized with alkyl chains *via* amide linkages at peripheries in order to utilise three intramolecular interactions, namely (i) π–π interactions of aromatic TPE cores, (ii) long alkyl chains on the periphery of TPE-core, which are designed to optimize the dispersive and van der Waals interactions, and (iii) hydrogen-bonding *via* amide functional

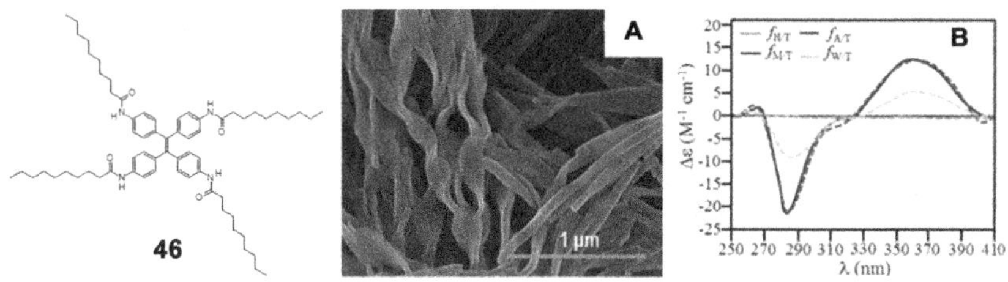

Fig. 10 (A) SEM images showing the formation process of the twisted-shaped microstructure deposited by solvent evaporation from **46** solutions in THF/ACN. (B) Circular dichroism spectra of **46** (alkyl-TPE = 10^{-5} M) in hexane/THF (1 : 9, v/v), MeOH/THF (1 : 9, v/v), ACN/THF (1 : 9, v/v), and water/THF (1 : 7, v/v). Adapted with permission from ref. 62, Anuradha, D.D. La, M. Al Kobaisi and S.V. Bhosale, Right handed chiral superstructures from achiral molecules: self-assembly with a twist, *Sci. Rep.*, 2015, **5**, 15652. Reproduced under Creative Commons license: https://creativecommons.org/licenses/by/4.0/.

groups, for the formation of controlled twisted chiral assembly. These interactions prevent crystallization and favour arrangement with the directional growth of twisted superstructure. The resultant achiral compounds displayed a typical AIE effect which was non emissive in the dilute THF solution with Φ_F of approximately 0.07% and strongly emissive in aggregate states with Φ_F of 19.7%, 19.6% and 9.8% in mixed solutions of water/THF, methanol/THF and hexane/THF, respectively. Furthermore, the self-assembly of alkyl-TPE were carried out in THF/MeOH and THF/acetonitrile mixtures, which resulted in the formation of ordered well-organized right-handed helical ribbons, 145 nm in width, 40 nm in thickness and tens of micrometers in length. The handedness of the twisted aggregates was confirmed by circular dichroism (CD) spectroscopy, which showed a characteristic of right-handed helical assembly for the TPE chromophores with a positive signal at 362 nm and a negative signal at 286 nm in THF/acetonitrile mixtures.

7 Applications

The TPE-based AIE-active luminogens possess several properties which can be utilised for sensing, templating, optoelectronic, mechanochromism, biological imaging,[63] solar energy harvesting[64] and many other applications. The properties associated with TPE core are its tendency for self-assembly, the AIE effect, and the possibility of conjugation of this moiety to others while keeping these essential properties unaffected, thus giving synthetic chemists a versatile building block to be added to their toolbox. The self-assembly of TPE and TPE-based molecular structures are driven by its strength of intermolecular interactions and solvophobic effects which in turn control the AIE effect. Taking advantage of TPE derivatives' properties, many applications are proposed, and herein, we give an overview of few very interesting examples.

7.1 Organic Light Emitting Diodes (OLEDs)

TPE-based AIE luminogens with controllable chemical, photophysical and functional properties in the solid state have been widely investigated for OLED applications. A typical configuration of a multilayer OLED includes ITO/NPB/TPE-based luminogens/TPBi/Alq_3/LiF/Al, in which NPB [(4,4-bis(1-naphthylphenylamino)biphenyl)] serves as a hole-transporting layer, TPBi is (2,2′,2″-(1,3,5-benzinetriyl)tris(1-phenyl-1-*H*-benzimidazole)), and Alq_3 acts as an electron-transporting layer. A range of different colour OLED emitters with high quantum yield and good stability can be fabricated from various TPE derivatives.

TPE core alone (**1**) gives a deep blue EL emission peak at 445 nm, which is similar to PL in its solid state. The device performance made of TPE is relatively low with maximum luminance (L_{max}) of merely 1800 cd m^{-2}, current efficiency (CE_{max}) of 0.45 cd A^{-1} and external quantum efficiency (EQE_{max}) of only 0.4%. Most of OLEDs fabricated from TPE-based AIE luminogens also display blue colour such as **1**,[65] **47**,[66] **48**,[10] **49**,[67] **50**,[68] **51**,[69] **52**,[68] and **53**[48] (Scheme 9). An outstanding blue OLED emitter was fabricated using TPATPE (**47**) as a light-emitting layer, which is strongly emissive at 492 nm with L_{max} up to 15 480 cd m^{-2}, CE_{max}, maximum power efficiency (PE_{max}) and EQE_{max} reach to 8.6 cd A^{-1}, 5.3 lm W^{-1} and 3.4%, respectively. However, one of the best blue OLED performances reported so far was observed from devices that were assembled using TTPEPy molecules (**49**). Thanks to its four TPE moieties, TTPEPy-based emitter gives the best EL performance. A blue light starts to emit at a lower voltage of 3.6 V at a wavelength of 490 nm with a L_{max} up to 36 300 cd m^{-2}, CE_{max} of 12.3 cd A^{-1} and EQE_{max} up to 4.95%.

Many green-light EL emitters were fabricated from various TPE-based luminogens as shown in Scheme 9. For instance, based on outstanding fluorescent efficiency in solid state and high carrier mobility an OLED device made of DPDPyE (**54**) exhibit a green emission at 516 nm.[70] The device can be turned on at very low voltage of 3.2 V with a L_{max} of 49 830 cd m^{-2} at 15V. The CE_{max} and EQE_{max} of this emitter are 10.2 cd A^{-1} and 3.3%, respectively, which is superior to a well-known green EL device fabricated using Alq_3. The best green EL OLED performance of this kind was also obtained from devices constructed using 3TPATPE **56**.[48] The device radiates a strong green emission peak at 514 nm, at a low voltage of 3.2 V, with L_{max}, CE_{max}, PE_{max} and EQE_{max} up to 33 700 cd m^{-2}, 13 cd A^{-1}, 11 lm W^{-1} and 4.4%, respectively. Other green light OLEDs were fabricated using TPE-based compounds **55**,[66] **57**,[71] and **58**[72] (Scheme 9).

Apart from dominant blue and green OLEDs in this field, few articles report devices with emission in different colours such as yellow, orange, red and white using TPE derivatives as a light-emitting layer (Scheme 10). For example, Xu *et al.*[73] reported a star-shaped red luminogen (T_2BT_2) (**59**) containing a benzo-2,1,3-thiadiazole core, TPA bridges, and TPE units. The OLED obtained from T_2BT_2 shows red light luminescence with a L_{max} of 13 535 cd m^{-2} and a CE_{max} of 6.81 cd A^{-1}. Another red light OLED made of TTPEBTTD (**60**) was reported by Zhao and co-workers.[63] The device radiates intense red light at 650 nm with L_{max}, E_{max} and EQE_{max} of 3750 cd m^{-2}, 2.4 cd A^{-1} and 3.7%, respectively. Zhao *et al.* reported a series of TPE-based luminogens among which an orange light emitting molecule with benzo-2,1,3-thiadiazole and thiophene linker units BTPETTD (**61**) is synthesized, and used as a light-emitting layer to fabricate orange OLED.[71] The device of BTPETTD emits orange-red light

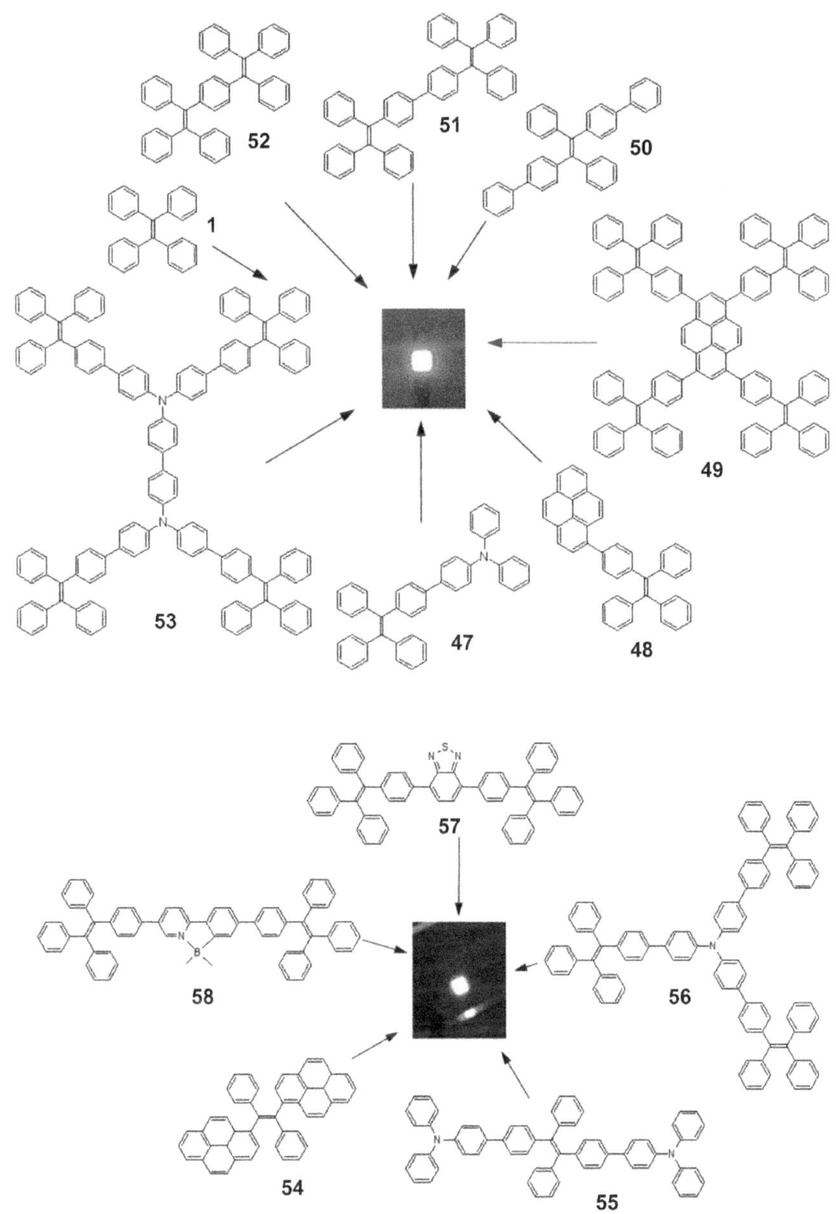

Scheme 9 Chemical structures of AIE-active luminogens showing bright light in OLED applications.

at the wavelength of 592 nm with a L_{max} of 8330 cd m^{-2}, E_{max} of 6.4 cd A^{-1} and EQE$_{\text{max}}$ of 3.1%.

7.2 Biological applications of TPE-based luminogens

Designing fluorogens for biological applications, usually imaging luminoges and probes to monitor a biological process or biological agents, require detailed understanding of the probe behaviour and the subject of

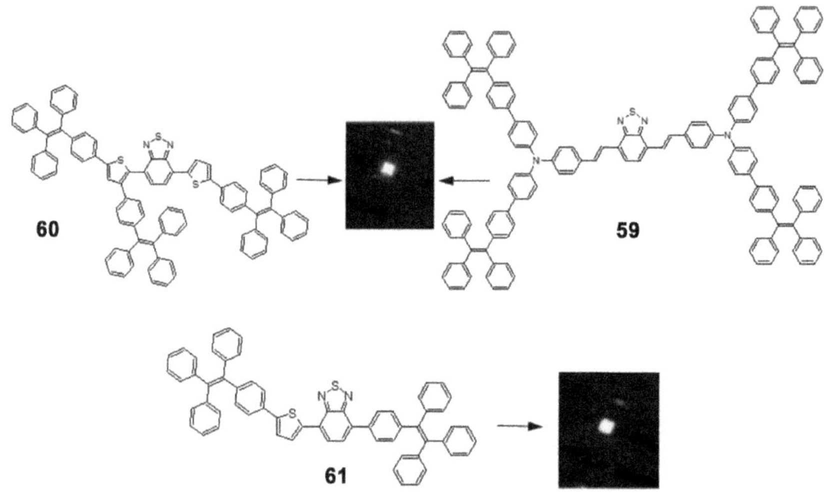

Scheme 10 Chemical structures of more AIE-active luminogens showing bright light in OLED applications.

interest. Although some luminogenic molecules can be used in their solubilised state where only the pH or a small ion to be monitored, but the luminescence signal is required when the probe is associated with a target analyte or aggregated in a self-assembly. The majority of probes fail the latter category due to the troublesome ACQ effect, where molecules in aggregates undergo short-range interactions that provides other pathways for excited molecules relaxation, resulting in quenching of the fluorescent emission. TPE derivatives *via* AIE effect offer a way to engineer new bioprobe molecules. For instance, for real-time monitoring of cell apoptosis, caspases can be used as the analyte to be monitored as the process signature. Shi *et al.*[74] designed a TPE based luminogen bioprobe composed of TPE moiety and a hydrophilic DEVD peptide sequence (**62**), which is almost non-emissive but displays strong fluorescence signal after the cleavage of the DEVD moieties in presence of caspase-3, which are activated in the apoptotic process. The (**62**) probe is highly soluble in aqueous solution, but cleaving the hydrophilic part of DEVD due to caspase-3, the residual hydrophobic TPE part forms aggregates and induces high emission (Fig. 11). This probe is capable of selective detection of caspase-3 activity and concentration in both solution and living cells as a function of fluorescent intensities. Earlier similar improved probe was discussed (**41**) where GFFY peptide sequence was conjugated to TPE to enhance the self-assembly and aggregation after the DEVD cleavage. The detection limit for this bioprobe is 0.54 pM caspase-3.

Tong *et al.*[75] designed, synthesised and investigated the multiresponsive behaviour of AIE active TPE derivative decorated with the electron-withdrawing 1,3-indan-dione group (IND-TPE) (**63**). This compound exhibited strong solvatochromism due to intramolecular charge transfer transition from TPE to IND moieties, giving redshift from apolar (green FL) to highly polar (red-orange FL) solvents. IND-TPE **63** displayed mechanochromic properties as well, where the as-prepared material emits green fluorescence

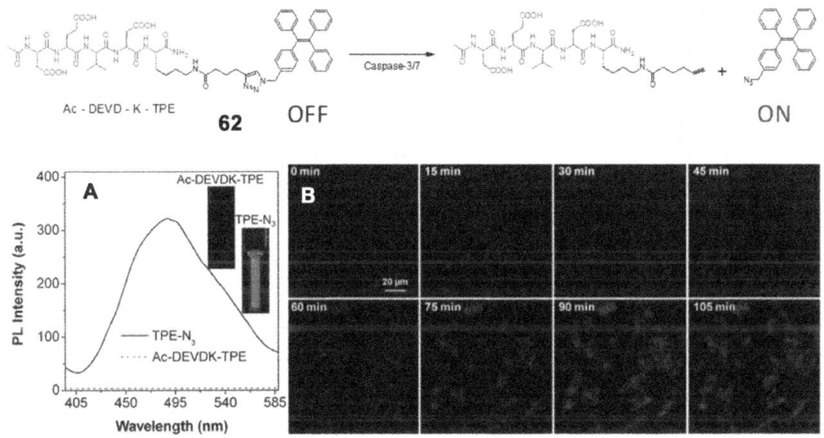

Fig. 11 Illustration of Ac-DEVDK-TPE (**62**) used as caspase activities monitor, and (A) Photoluminescence (PL) spectra of TPE-N3 and **62** in DMSO/water (v/v = 1 : 199), inset is photographs of TPE-N3 and **62** in DMSO/water (v/v = 1 : 199) taken under illumination by an UV lamp. (B) Real-time fluorescence images showing the cell apoptotic process of MCF-7 cells with **62** at room temperature. STS (1 μM) was used to induce cell apoptosis. Adapted with permission from ref. 74. Copyright 2012 American Chemical Society.

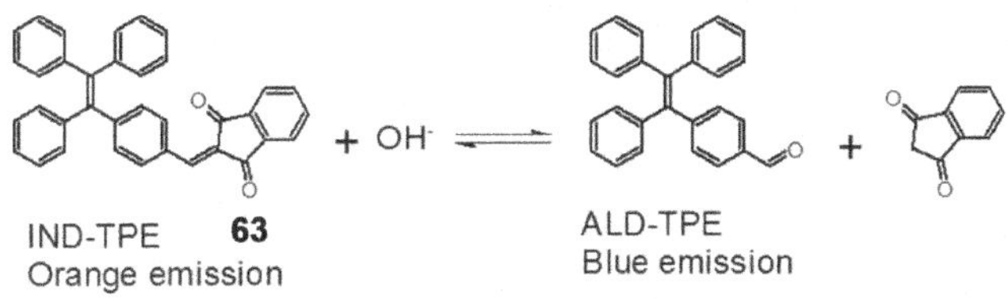

Scheme 11 Schematic illustration of the principle of dual fluorescent response of IND-TPE (**63**) to hydroxyl anion.

and the ground sample emits orange, which can be reversed by thermal annealing and solvent fuming. The molecule **63** was very sensitive to pH, giving aldehyde modified TPE with blue emission *via* hydrolysis reaction (Scheme 11). Tong *et al.* in one demonstration showed a selective signal produced by **63** towards arginine (Arg) and lysine (Lys) among other amino acids. They assigned this selective response to the basic nature of these amino acids with isoelectric point (pI) values of 10.76 and 9.74, respectively. Using this probe, only Arg and Lys exhibited red-orange FL quenching among 16 amino acids with the detection limit of about 100 μM.

Another bioprobe designed and synthesised by Zhao *et al.*[76] demonstrated a selectivity towards a mitochondrion in a living cell which require positively charged lipophilic stains. This bioprobe is TPE core conjugated to pyridinium unit through vinyl functionality (TPE-Py) (**64**) giving a biocompatible yellow-emissive AIE dye. A standard cell-staining protocol was employed to image HeLa cells using TPE-Py **64**, where nano-aggregates were prepared in the essential medium-DMSO mixture. As a result, a distinct yellow emission was observed, which was ascribed to the interaction between **64** and mitochondria in the living cell (Fig. 12A).

In another example, TPE used to decorate the ACQ porphyrin producing an AIE active bioprobe (PorTPE) (**65**) which was utilized in cellular

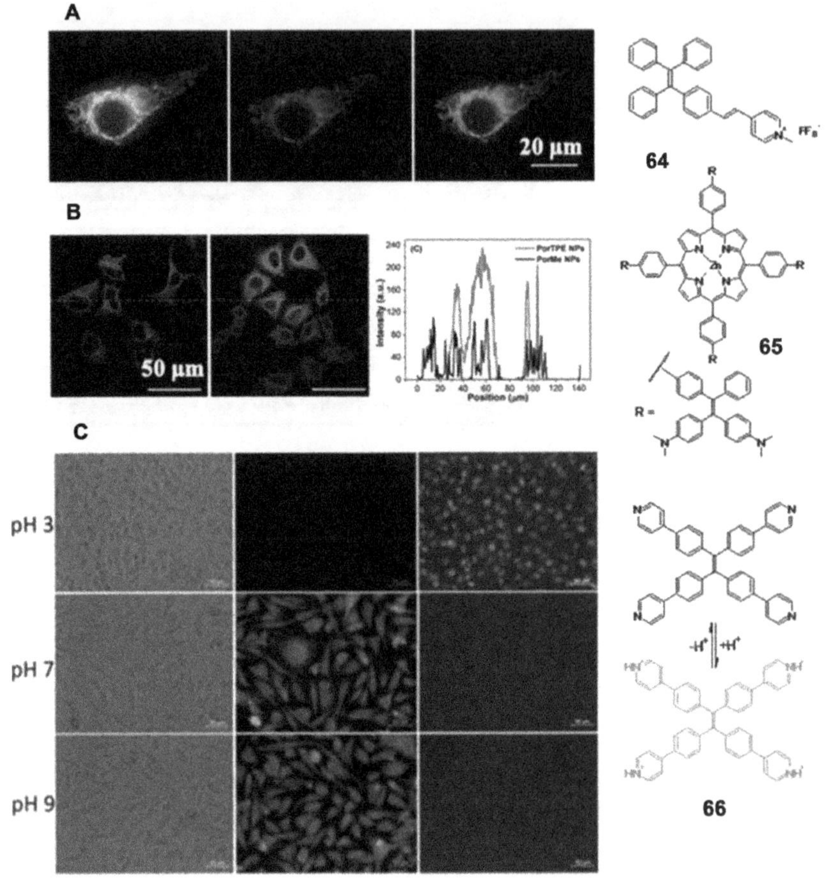

Fig. 12 (A) Fluorescent images of a HeLa cell incubated with TPE-Py (**64**) (5 mM) (left, yellow fluorescence), MitoTracker Red (100 nM) for 15 min at 37 °C (centre, red fluorescence), and their merged picture (right). (B) Confocal images of fixed HeLa cells after incubation with 2 mM PorMe NPs or PorTPE NPs (**65**), and fluorescence intensity profiles of cell images are indicated by the white dashed line. (C) Fluorescence microscopic images of PC-3 cells treated with 5 mg mL^{-1} of **66** for two hours at different pH conditions. At lower pH cells show fluorescence in the green channel while at pHs higher than 7 the cells show fluorescence in the blue channel, and no fluorescence was observed in the red channel. Control untreated cells did not show any fluorescence. In the left panel bright field phase, contrast images are shown for the same field of view. A is adapted from ref. 76 with permission from the Royal Society of Chemistry, B is adapted from ref. 42 with permission from the Royal Society of Chemistry and C is adapted from ref. 77 with permission from the Royal Society of Chemistry.

imaging.[42] The aggregates of PorTPE nanoparticles in HeLa cells show excellent and distinct brightness in red color, which is superior to the bioimaging performance of well-identified zinc tetramethylphenyl porphyrin dye (Fig. 12B). Using a similar approach we reported an efficient dye (**66**) for imaging human prostate cancer (PC-3) living cells.[77] This bioprobe was made by substituting TPE with pyridyl groups giving tetrapyridyl-substituted TPE (**66**). The fluorescent microscopy as shown in Fig. 12C, which shows a bright blue emission peak between 450 and 490 nm under neutral and alkaline pH conditions, undergoes a significant red shift in acidic conditions giving green emission in the range of

520–560 nm. This bright image can be obtained at a low dye concentration of 5 μg mL^{-1}.

7.3 Sensing applications

In addition to the vast potential in molecular design and application of TPE derivative as bioprobes using its photo-physical properties and self-assembling tendencies, they can be utilized in designing small ions and molecules sensors. For example, we recently reported the design and synthesis of tetraamino-tetraphenylethene (TA-TPE) (**67**), which could be used as a sensitive probe for nitrite ions through a visible colour change in aqueous condition.[78] The TA-TPE is colourless in aqueous solution. However, in the presence of NO_2^- ions, it quickly changes to dark yellow with detection limit as low as 17.7 ppb. Fig. 13A presents the visual excellent selectivity of TA-TPE toward NO_2^- ions.

In another example a selective Cu(ii) detector was developed using an AIE active TPE macrocycle nanofibers **26** (Fig. 13B).[45] In the absence of copper ions, no absorption peak was observed at about 450 nm for the solution of TPE macrocycle. However, a strong absorption was observed at this wavelength upon copper addition. This drastic alternation of UV–Vis spectrum with the presence of copper ions can be even visualized at Cu(ii) concentration as low as 1.0×10^{-5} M, which is able to determine the level of copper ions that exceed the safety level of copper ions in drinking water, which is 2.0×10^{-5} M according to Environmental Protection Agency (EPA). Interestingly, there was no obvious change in the UV–Vis spectra when other metal ions, such as Fe^{2+}, Co^{2+}, Ni^{2+}, Zn^{2+}, Cd^{2+}, Hg^{2+}, Mn^{2+}, Ag^+, Cr^{3+}, Pb^{2+}, Al^{3+}, Ca^{2+}, Mg^{2+}, K^+ and Na^+, were added to the solution of **26**. These results indicate that aggregated TPE macrocycle nanofibers are a great candidate for highly sensitive and selective detection of copper ions in drinking water as well as in living tissue-containing aqueous media.

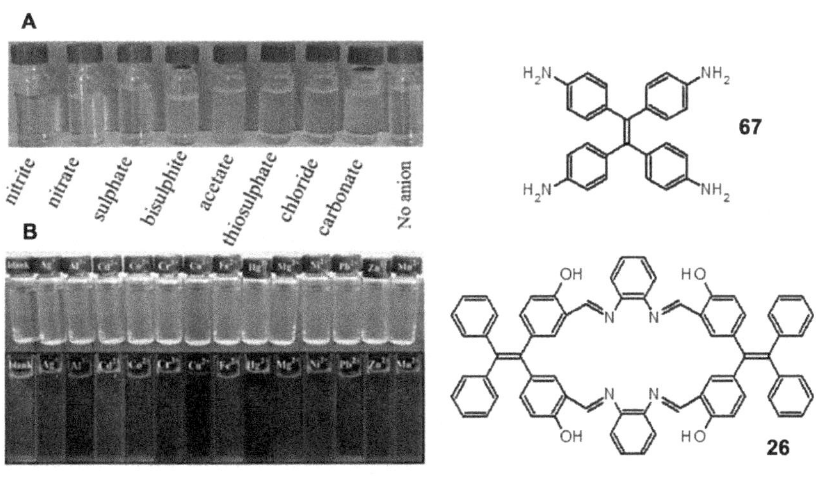

Fig. 13 (A) Colorimetric response of TA-TPE (**67**) toward various competitive anions. (B) Photos of **26** in H_2O – THF 2 : 1 under daylight (Top) and photos of **26** in H_2O – THF 9 : 1 under 365 nm light after addition of the metal ion ([**26**] = 1.0×10^{-5} M, [metal] = 2.0×10^{-5} M) (Bottom). A is adapted from ref. 78 with permission from the Royal Society of Chemistry and B is adapted from ref. 45 with permission from the Royal Society of Chemistry.

7.4 Photovoltaic cells

Taking advantage of strong absorption in the visible region, diversity of nanostructure and facile tunability in electronic property, TPE derivatives are used as non-fullerene acceptors for bulk heterojunction organic solar cells (OSCs). The designed and prepared TPE derivatives also need to have strong electron mobility, good solubility, and thermal stability. Considering these properties, a TPE core-based 3D structure small molecule acceptor was synthesized, and employed for photovoltaic devices. The molecular design of this acceptor included a tetra-PDI substitution of the central TPE core. This acceptor exhibited a weak aggregation and forms a smooth and amorphous film with good electron-transport ability and a high LUMO level.[79] A bulk heterojunction device was fabricated from this 3D molecule and showed an excellent power-conversion efficiency (PCE) of 5.53%. Recently, we also reported the design and synthesis of a TPE-based electron acceptor with diketopyrrolopyrrole (DPP) terminals. This acceptor revealed good solubility, thermal stability, and high LUMO level. An organic solar cell based on this molecule exhibited high conversion efficiency, up to 3.86%, with an open-circuit voltage of 1.18 V.[64]

In summary, within last ten years, several research groups have used TPE core for creating various functional architectures. Herein we shed some light in strategies to utilise TPE molecular structure and properties in order to design TPE-based functional architectures to fulfil needs in the areas of sensing, in particular biosensing, light harvesting, optoelectronics and guided self-assembly. In addition, we covered synthetic strategies in order to control and tune TPE derivatives' photo-physical and chemical behaviours *via* modification at the periphery and conjugation to biological moieties such as peptides. Furthermore, such modifications can lead TPE derivative to assume preferred conformations to construct supramolecular systems with a wide variety of soft matter with applications in biomedical and sensing research areas. Such supramolecular materials in combination with AIE properties of TPE derivatives offered researchers an open field to investigate molecular switching through mechanoluminescence. In conclusion, we revealed that TPE represents itself as an ideal building block to construct diverse functional materials for the needs of modern humanity.

Acknowledgements

D.D.L. thanks RMIT University for financial support. S.V.B. acknowledges the Australian Research Council under a Future Fellowship Scheme (FT110100152).

References

1 R. Hu, N. L. Leung and B. Z. Tang, AIE macromolecules: syntheses, structures and functionalities, *Chem. Soc. Rev.*, 2014, **43**(13), 4494–4562.

2 J. Huang, N. Sun, Y. Dong, R. Tang, P. Lu, P. Cai, Q. Li, D. Ma, J. Qin and Z. Li, Similar or Totally Different: The Control of Conjugation Degree through Minor Structural Modifications, and Deep-Blue Aggregation-Induced Emission Luminogens for Non-Doped OLEDs, *Adv. Funct. Mater.*, 2013, **23**(18), 2329–2337.

3 J. Luo, Z. Xie, J. W. Lam, L. Cheng, H. Chen, C. Qiu, H. S. Kwok, X. Zhan, Y. Liu and D. Zhu, Aggregation-induced emission of 1-methyl-1, 2, 3, 4, 5-pentaphenylsilole, *Chem. Commun.*, 2001, **18**, 1740–1741.
4 G. Qian, Z. Zhong, M. Luo, D. Yu, Z. Zhang, Z. Y. Wang and D. Ma, Simple and Efficient Near-Infrared Organic Chromophores for Light-Emitting Diodes with Single Electroluminescent Emission above 1000 nm, *Adv. Mater.*, 2009, **21**(1), 111–116.
5 C. Zhang, S. Jin, K. Yang, X. Xue, Z. Li, Y. Jiang, W.-Q. Chen, L. Dai, G. Zou and X.-J. Liang, Cell membrane tracker based on restriction of intramolecular rotation, *ACS Appl. Mater. Interfaces*, 2014, **6**(12), 8971–8975.
6 G.-F. Zhang, Z.-Q. Chen, M. P. Aldred, Z. Hu, T. Chen, Z. Huang, X. Meng and M.-Q. Zhu, Direct validation of the restriction of intramolecular rotation hypothesis via the synthesis of novel ortho-methyl substituted tetraphenylethenes and their application in cell imaging, *Chem. Commun.*, 2014, **50**(81), 12058–12060.
7 C. L. Chiang, S. M. Tseng, C. T. Chen, C. P. Hsu and C. F. Shu, Influence of molecular dipoles on the photoluminescence and electroluminescence of dipolar spirobifluorenes, *Adv. Funct. Mater.*, 2008, **18**(2), 248–257.
8 B. S. Gaylord, S. Wang, A. J. Heeger and G. C. Bazan, Water-soluble conjugated oligomers: effect of chain length and aggregation on photoluminescence-quenching efficiencies, *J. Am. Chem. Soc.*, 2001, **123**(26), 6417–6418.
9 D. Ding, K. Li, B. Liu and B. Z. Tang, Bioprobes based on AIE fluorogens, *Acc. Chem. Res.*, 2013, **46**(11), 2441–2453.
10 Z. Zhao, P. Lu, J. W. Lam, Z. Wang, C. Y. Chan, H. H. Sung, I. D. Williams, Y. Ma and B. Z. Tang, Molecular anchors in the solid state: Restriction of intramolecular rotation boosts emission efficiency of luminogen aggregates to unity, *Chem. Sci.*, 2011, **2**(4), 672–675.
11 G. Sun, Y. Zhao and W. Liang, Aggregation-Induced Emission Mechanism of Dimethoxy-Tetraphenylethylene in Water Solution: Molecular Dynamics and QM/MM Investigations, *J. Chem. Theory Comput.*, 2015, **11**(5), 2257–2267.
12 P. de Boissieu, New Method for the Preparation of Tetraphenylethylene, *J. Chem. Soc., Abstr.*, 1888, **54**, 928–972.
13 W. Wang, T. Lin, M. Wang, T.-X. Liu, L. Ren, D. Chen and S. Huang, Aggregation emission properties of oligomers based on tetraphenylethylene, *J. Phys. Chem. B*, 2010, **114**(18), 5983–5988.
14 A. Schreivogel, J. Maurer, R. Winter, A. Baro and S. Laschat, Synthesis and electrochemical properties of tetrasubstituted tetraphenylethenes, *Eur. J. Org. Chem.*, 2006, **2006**(15), 3395–3404.
15 G. Hua, Y. Li, A. M. Slawin and J. D. Woollins, Stereoselective synthesis of olefins by a reductive coupling reaction, *Dalton Trans.*, 2007, **15**, 1477–1480.
16 L. Zhang, X. Yu, L. Zhang, X. Zhou and Y. Lin, A simple and practical catalytic electron transfer system composed of TiCl 4 and metallic Yb: application in carbonyl olefination and insight into the mechanism, *Org. Chem. Front.*, 2014, **1**(8), 929–935.
17 S. Nadri, M. Joshaghani and E. Rafiee, Catalytic Performance of a Phospha-palladacycle Bearing a Biphenyl Moiety, Possessing an sp3 C– Pd Bond, toward the Heck Coupling Reaction, *Organometallics*, 2009, **28**(21), 6281–6287.
18 K. Okuma, K. Kojima, K. Oyama, K. Kubo and K. Shioji, Selenobenzophenones and Diazoalkanes: Isolation of Tetraarylethylenes by the Reaction of Benzophenone Hydrazones with Diselenium Dibromide, *Eur. J. Org. Chem.*, 2004, **2004**(4), 820–825.
19 Y. Tezuka, A. Hashimoto, K. Ushizaka and K. Imai, Generation and reactions of novel copper carbenoids through a stoichiometric reaction of copper

metal with gem-dichlorides in dimethyl sulfoxide, *J. Org. Chem.*, 1990, **55**(1), 329–333.
20 K. Sajna, V. Srinivas and K. Kumara Swamy, Efficient Palladium-Catalyzed Double Arylation of Phosphonoalkynes and Diarylalkynes in Water: Use of a Dinuclear Palladium (I) Catalyst, *Adv. Synth. Catal.*, 2010, **352**(17), 3069–3081.
21 N. Goswami, Q. Yao, Z. Luo, J. Li, T. Chen and J. Xie, Luminescent Metal Nanoclusters with Aggregation-Induced Emission, *J. Phys. Chem. Lett.*, 2016, **7**(6), 962–975.
22 V. V. Roznyatovskiy, R. Carmieli, S. M. Dyar, K. E. Brown and M. R. Wasielewski, Photodriven Charge Separation and Transport in Self-Assembled Zinc Tetrabenzotetraphenylporphyrin and Perylenediimide Charge Conduits, *Angew. Chem., Int. Ed.*, 2014, **53**(13), 3457–3461.
23 B. Rybtchinski, L. E. Sinks and M. R. Wasielewski, Combining light-harvesting and charge separation in a self-assembled artificial photosynthetic system based on perylenediimide chromophores, *J. Am. Chem. Soc.*, 2004, **126**(39), 12268–12269.
24 S. Fabiano, S. Braun, M. Fahlman, X. Crispin and M. Berggren, Effect of Gate Electrode Work-Function on Source Charge Injection in Electrolyte-Gated Organic Field-Effect Transistors, *Adv. Funct. Mater.*, 2014, **24**(5), 695–700.
25 X. Zhang, S. Rehm, M. M. Safont-Sempere and F. Würthner, Vesicular perylene dye nanocapsules as supramolecular fluorescent pH sensor systems, *Nat. Chem.*, 2009, **1**(8), 623–629.
26 Q. Zhao, S. Zhang, Y. Liu, J. Mei, S. Chen, P. Lu, A. Qin, Y. Ma, J. Z. Sun and B. Z. Tang, Tetraphenylethenyl-modified perylene bisimide: aggregation-induced red emission, electrochemical properties and ordered microstructures, *J. Mater. Chem.*, 2012, **22**(15), 7387–7394.
27 M. P. Aldred, G.-F. Zhang, C. Li, G. Chen, T. Chen and M.-Q. Zhu, Optical properties and red to near infrared piezo-responsive fluorescence of a tetraphenylethene–perylenebisimide–tetraphenylethene triad, *J. Mater. Chem. C*, 2013, **1**(40), 6709–6718.
28 Q. Zhao, X. A. Zhang, Q. Wei, J. Wang, X. Y. Shen, A. Qin, J. Z. Sun and B. Z. Tang, Tetraphenylethene modified perylene bisimide: effect of the number of substituents on AIE performance, *Chem. Commun.*, 2012, **48**(95), 11671–11673.
29 N.-H. Xie, C. Li, J.-X. Liu, W.-L. Gong, B. Z. Tang, G. Li and M.-Q. Zhu, The synthesis and aggregation-induced near-infrared emission of terrylenediimide–tetraphenylethene dyads, *Chem. Commun.*, 2016, **52**(34), 5808–5811.
30 Y. J. Wang, Z. Li, J. Tong, X. Y. Shen, A. Qin, J. Z. Sun and B. Z. Tang, The fluorescence properties and aggregation behavior of tetraphenylethene–perylenebisimide dyads, *J. Mater. Chem. C*, 2015, **3**(15), 3559–3568.
31 O. Yushchenko, R. V. Hangarge, S. Mosquera-Vazquez, S. V. Bhosale and E. Vauthey, Electron, Hole, Singlet, and Triplet Energy Transfer in Photoexcited Porphyrin-Naphthalenediimide Dyads, *J. Phys. Chem. B*, 2014, **119**(24), 7308–7320.
32 M. Al Kobaisi, S. V. Bhosale, K. Latham and S. V. Bhosale, Functional Naphthalene Diimides: Synthesis, Properties and Applications, *Chem. Rev.*, 2015, DOI: 10.1021/acs.chemrev.6b00160.
33 H. Katz, A. Lovinger, J. Johnson, C. Kloc, T. Siegrist, W. Li, Y.-Y. Lin and A. Dodabalapur, A soluble and air-stable organic semiconductor with high electron mobility, *Nature*, 2000, **404**(6777), 478–481.
34 A. Rananaware, D. D. La and S. V. Bhosale, Solvophobic control aggregation-induced emission of tetraphenylethene-substituted naphthalene diimide via intramolecular charge transfer, *RSC Adv.*, 2015, **5**(77), 63130–63134.

35 A. Rananaware, D. D. La, S. M. Jackson and S. V. Bhosale, Construction of a highly efficient near-IR solid emitter based on naphthalene diimide with AIE-active tetraphenylethene periphery, *RSC Adv.*, 2016, **6**(20), 16250–16255.
36 B. Albinsson and J. Mårtensson, Long-range electron and excitation energy transfer in donor–bridge–acceptor systems, *J. Photochem. Photobiol., C*, 2008, **9**(3), 138–155.
37 C. Herrero, B. Lassalle-Kaiser, W. Leibl, A. W. Rutherford and A. Aukauloo, Artificial systems related to light driven electron transfer processes in PSII, *Coord. Chem. Rev.*, 2008, **252**(3), 456–468.
38 S. Singh, A. Aggarwal, N. D. K. Bhupathiraju, K. Tiwari and C. M. Drain, Glycosylated Porphyrins, Phthalocyanines, and Other Porphyrinoids for Diagnostics and Therapeutics, *Chem. Rev.*, 2015, **115**(18), 10261–10306.
39 Q. T. Nguyen and R. Y. Tsien, *Fluorescence-guided surgery with live molecula*r navigation [mdash] a new cutting edge, *Nat. Rev. Cancer*, 2013, **13**(9), 653–662.
40 N. Aratani, D. Kim and A. Osuka, Discrete cyclic porphyrin arrays as artificial light-harvesting antenna, *Acc. Chem. Res.*, 2009, **42**(12), 1922–1934.
41 A. Rananaware, R. S. Bhosale, K. Ohkubo, H. Patil, L. A. Jones, S. L. Jackson, S. Fukuzumi, S. V. Bhosale and S. V. Bhosale, Tetraphenylethene-Based Star Shaped Porphyrins: Synthesis, Self-assembly, and Optical and Photophysical Study, *J. Org. Chem.*, 2015, **80**(8), 3832–3840.
42 B. Guo, X. Cai, S. Xu, S. M. A. Fateminia, J. Liu, J. Liang, G. Feng, W. Wu and B. Liu, Decoration of porphyrin with tetraphenylethene: converting a fluorophore with aggregation-caused quenching to aggregation-induced emission enhancement, *J. Mater. Chem. B*, 2016, **4**(27), 4690–4695.
43 S. Song and Y.-S. Zheng, Hollow Spheres Self-Assembled by a Tetraphenylethylene Macrocycle and Their Transformation to Bird Nests under Ultrasound, *Org. Lett.*, 2013, **15**(4), 820–823.
44 S. Song, H.-F. Zheng, H.-T. Feng and Y.-S. Zheng, Microtubes and hollow microspheres formed by winding of nanoribbons from self-assembly of tetraphenylethylene amide macrocycles, *Chem. Commun.*, 2014, **50**(96), 15212–15215.
45 H.-T. Feng, S. Song, Y.-C. Chen, C.-H. Shen and Y.-S. Zheng, Self-assembled tetraphenylethylene macrocycle nanofibrous materials for the visual detection of copper (II) in water, *J. Mater. Chem. C*, 2014, **2**(13), 2353–2359.
46 A. Rananaware, D. D. La and S. V. Bhosale, Aggregation-induced emission of a star-shape luminogen based on cyclohexanehexone substituted with AIE active tetraphenylethene functionality, *RSC Adv.*, 2015, **5**(69), 56270–56273.
47 J. Wang, J. Mei, R. Hu, J. Z. Sun, A. Qin and B. Z. Tang, Click synthesis, aggregation-induced emission, E/Z isomerization, self-organization, and multiple chromisms of pure stereoisomers of a tetraphenylethene-cored luminogen, *J. Am. Chem. Soc.*, 2012, **134**(24), 9956–9966.
48 W. Z. Yuan, P. Lu, S. Chen, J. W. Lam, Z. Wang, Y. Liu, H. S. Kwok, Y. Ma and B. Z. Tang, Changing the Behavior of Chromophores from Aggregation-Caused Quenching to Aggregation-Induced Emission: Development of Highly Efficient Light Emitters in the Solid State, *Adv. Mater.*, 2010, **22**(19), 2159–2163.
49 K. Han and B.-K. Cho, Chain-dependent emission color codes of extended tetraphenylethylene derivatives: discrimination between water and methanol, *RSC Adv.*, 2015, **5**(13), 9510–9517.
50 X. Yan, M. Wang, T. R. Cook, M. Zhang, M. L. Saha, Z. Zhou, X. Li, F. Huang and P. J. Stang, Light-Emitting Superstructures with Anion Effect: Coordination-Driven Self-Assembly of Pure Tetraphenylethylene Metallacycles and Metallacages, *J. Am. Chem. Soc.*, 2016, **138**(13), 4580–4588.

51 H. Engelkamp and S. Middelbeek, Self-assembly of disk-shaped molecules to coiled-coil aggregates with tunable helicity, *Science*, 1999, **284**(5415), 785–788.
52 M. A. Mateos-Timoneda, M. Crego-Calama and D. N. Reinhoudt, Supramolecular chirality of self-assembled systems in solution, *Chem. Soc. Rev.*, 2004, **33**(6), 363–372.
53 C. Dressel, T. Reppe, M. Prehm, M. Brautzsch and C. Tschierske, Chiral self-sorting and amplification in isotropic liquids of achiral molecules, *Nat. Chem.*, 2014, **6**(11), 971–977.
54 H. Li, J. Cheng, Y. Zhao, J. W. Lam, K. S. Wong, H. Wu, B. S. Li and B. Z. Tang, L-Valine methyl ester-containing tetraphenylethene: aggregation-induced emission, aggregation-induced circular dichroism, circularly polarized luminescence, and helical self-assembly, *Mater. Horiz.*, 2014, **1**(5), 518–521.
55 H. Li, J. Cheng, H. Deng, E. Zhao, B. Shen, J. W. Lam, K. S. Wong, H. Wu, B. S. Li and B. Z. Tang, Aggregation-induced chirality, circularly polarized luminescence, and helical self-assembly of a leucine-containing AIE luminogen, *J. Mater. Chem. C*, 2015, **3**(10), 2399–2404.
56 H. Li, X. Zheng, H. Su, J. W. Y. Lam, K. Sing Wong, S. Xue, X. Huang, X. Huang, B. S. Li and B. Z. Tang, Synthesis, optical properties, and helical self-assembly of a bivaline-containing tetraphenylethene, *Sci. Rep.*, 2016, **6**, 19277.
57 W. Z. Yuan, F. Mahtab, Y. Gong, Z.-Q. Yu, P. Lu, Y. Tang, J. W. Lam, C. Zhu and B. Z. Tang, Synthesis and self-assembly of tetraphenylethene and biphenyl based AIE-active triazoles, *J. Mater. Chem.*, 2012, **22**(21), 10472–10479.
58 C. Zhang, C. Liu, X. Xue, X. Zhang, S. Huo, Y. Jiang, W.-Q. Chen, G. Zou and X.-J. Liang, Salt-responsive self-assembly of luminescent hydrogel with intrinsic gelation-enhanced emission, *ACS Appl. Mater. Interfaces*, 2013, **6**(2), 757–762.
59 M.-Y. Yeh, C.-W. Huang, J.-W. Chang, Y.-T. Huang, J.-H. Lin, S.-M. Hsu, S.-C. Hung and H.-C. Lin, A novel nanostructured supramolecular hydrogel self-assembled from tetraphenylethylene-capped dipeptides, *Soft Matter*, 2016, **12**(30), 6347–6351.
60 A. Han, H. Wang, R. T. Kwok, S. Ji, J. Li, D. Kong, B. Z. Tang, B. Liu, Z. Yang and D. Ding, Peptide-Induced AIEgen Self-Assembly: A New Strategy to Realize Highly Sensitive Fluorescent Light-Up Probes, *Anal. Chem.*, 2016, **88**(7), 3872–3878.
61 C. Zhang, Y. Li, X. Xue, P. Chu, C. Liu, K. Yang, Y. Jiang, W.-Q. Chen, G. Zou and X.-J. Liang, A smart pH-switchable luminescent hydrogel, *Chem. Commun.*, 2015, **51**(20), 4168–4171.
62 Anuradha, D. D. La, M. Al Kobaisi and S. V. Bhosale, Right handed chiral superstructures from achiral molecules: self-assembly with a twist, *Sci. Rep.*, 2015, **5**, 15652.
63 Z. Zhao, J. Geng, Z. Chang, S. Chen, C. Deng, T. Jiang, W. Qin, J. W. Lam, H. S. Kwok and H. Qiu, A tetraphenylethene-based red luminophor for an efficient non-doped electroluminescence device and cellular imaging, *J. Mater. Chem.*, 2012, **22**(22), 11018–11021.
64 A. Rananaware, A. Gupta, J. Li, A. Bilic, L. Jones, S. Bhargava and S. V. Bhosale, A four-directional non-fullerene acceptor based on tetraphenylethylene and diketopyrrolopyrrole functionalities for efficient photovoltaic devices with a high open-circuit voltage of 1.18 V, *Chem. Commun.*, 2016, **52**(55), 8522–8525.
65 Z. Zhao, S. Chen, J. W. Y. Lam, C. K. W. Jim, C. Y. K. Chan, Z. Wang, P. Lu, C. Deng, H. S. Kwok, Y. Ma and B. Z. Tang, Steric Hindrance, Electronic Communication, and Energy Transfer in the Photo- and Electroluminescence Processes of Aggregation-Induced Emission Luminogens, *J. Phys. Chem. C*, 2010, **114**(17), 7963–7972.

66 Y. Liu, S. Chen, J. W. Lam, P. Lu, R. T. Kwok, F. Mahtab, H. S. Kwok and B. Z. Tang, Tuning the electronic nature of aggregation-induced emission luminogens with enhanced hole-transporting property, *Chem. Mater.*, 2011, **23**(10), 2536–2544.
67 Z. Zhao, S. Chen, J. W. Lam, P. Lu, Y. Zhong, K. S. Wong, H. S. Kwok and B. Z. Tang, Creation of highly efficient solid emitter by decorating pyrene core with AIE-active tetraphenylethene peripheries, *Chem. Commun.*, 2010, **46**(13), 2221–2223.
68 Z. Zhao, J. W. Y. Lam and B. Z. Tang, Tetraphenylethene: a versatile AIE building block for the construction of efficient luminescent materials for organic light-emitting diodes, *J. Mater. Chem.*, 2012, **22**(45), 23726–23740.
69 Z. Zhao, S. Chen, X. Shen, F. Mahtab, Y. Yu, P. Lu, J. W. Lam, H. S. Kwok and B. Z. Tang, Aggregation-induced emission, self-assembly, and electroluminescence of 4, 4′-bis (1, 2, 2-triphenylvinyl) biphenyl, *Chem. Commun.*, 2010, **46**(5), 686–688.
70 Z. Zhao, S. Chen, J. W. Y. Lam, Z. Wang, P. Lu, F. Mahtab, H. H. Y. Sung, I. D. Williams, Y. Ma, H. S. Kwok and B. Z. Tang, Pyrene-substituted ethenes: aggregation-enhanced excimer emission and highly efficient electroluminescence, *J. Mater. Chem.*, 2011, **21**(20), 7210–7216.
71 Z. Zhao, C. Deng, S. Chen, J. W. Lam, W. Qin, P. Lu, Z. Wang, H. S. Kwok, Y. Ma and H. Qiu, Full emission color tuning in luminogens constructed from tetraphenylethene, benzo-2, 1, 3-thiadiazole and thiophene building blocks, *Chem. Commun.*, 2011, **47**(31), 8847–8849.
72 W. Z. Yuan, S. Chen, J. W. Y. Lam, C. Deng, P. Lu, H. H. Y. Sung, I. D. Williams, H. S. Kwok, Y. Zhang and B. Z. Tang, Towards high efficiency solid emitters with aggregation-induced emission and electron-transport characteristics, *Chem. Commun.*, 2011, **47**(40), 11216–11218.
73 H. Li, Z. Chi, X. Zhang, B. Xu, S. Liu, Y. Zhang and J. Xu, New thermally stable aggregation-induced emission enhancement compounds for non-doped red organic light-emitting diodes, *Chem. Commun.*, 2011, **47**(40), 11273–11275.
74 H. Shi, R. T. Kwok, J. Liu, B. Xing, B. Z. Tang and B. Liu, Real-time monitoring of cell apoptosis and drug screening using fluorescent light-up probe with aggregation-induced emission characteristics, *J. Am. Chem. Soc.*, 2012, **134**(43), 17972–17981.
75 J. Tong, Y. Wang, J. Mei, J. Wang, A. Qin, J. Z. Sun and B. Z. Tang, A 1, 3-Indandione-Functionalized Tetraphenylethene: Aggregation-Induced Emission, Solvatochromism, Mechanochromism, and Potential Application as a Multiresponsive Fluorescent Probe, *Chem. – Eur. J.*, 2014, **20**(16), 4661–4670.
76 N. Zhao, M. Li, Y. Yan, J. W. Lam, Y. L. Zhang, Y. S. Zhao, K. S. Wong and B. Z. Tang, A tetraphenylethene-substituted pyridinium salt with multiple functionalities: synthesis, stimuli-responsive emission, optical waveguide and specific mitochondrion imaging, *J. Mater. Chem. C*, 2013, **1**(31), 4640–4646.
77 A. Rananaware, R. S. Bhosale, H. Patil, M. Al Kobaisi, A. Abraham, R. Shukla, S. V. Bhosale and S. V. Bhosale, Precise aggregation-induced emission enhancement via H^+ sensing and its use in ratiometric detection of intracellular pH values, *RSC Adv.*, 2014, **4**(103), 59078–59082.
78 Anuradha, K. Latham and S. V. Bhosale, Selective detection of nitrite ion by an AIE-active tetraphenylethene dye through a reduction step in aqueous media, *RSC Adv.*, 2016, **6**(51), 45009–45013.
79 Y. Liu, C. Mu, K. Jiang, J. Zhao, Y. Li, L. Zhang, Z. Li, J. Y. L. Lai, H. Hu and T. Ma, A Tetraphenylethylene Core-Based 3D Structure Small Molecular Acceptor Enabling Efficient Non-Fullerene Organic Solar Cells, *Adv. Mater.*, 2015, **27**(6), 1015–1020.

A review of two-dimensional nanomaterials beyond graphene

Aleksander A. Tedstone,*[a] Jack R. Brent*[b] and David J. Lewis*[c]
DOI: 10.1039/9781782620358-00108

The principal interest in two-dimensional (2D) nanomaterials has been in their electronic properties, but as understanding of them has broadened, a wide range of applications have been explored. This chapter outlines recent literature concerning the synthesis, modification, and applications of two-dimensional nanomaterials beyond graphene. Recent theoretical propositions of these novel two-dimensional nanomaterials and their applications are also discussed.

1 Introduction

Nanomaterials with one or more dimension of nanometre length scale (1–100 nm) have been of growing interest to the scientific community due to their intrinsic properties which arise from physical confinement of materials at a miniscule length scale and the socio-economic driver of the prospect of ever-more miniaturised devices. The principal interest in two-dimensional (2D) nanomaterials has been in their electronic properties, but as understanding of them has broadened, a wide range of applications have been explored. For instance, their extremely high surface areas are beneficial for catalysis, an application that has been demonstrated with a number of 2D materials,[1–4] and is likely to continue evolving as the library of materials expands and their synthesis becomes more refined. Biomedical applications of many types of nanomaterials have been the subject of intensive research, and 2D materials are no exception, finding use in imaging, drug delivery and targeted therapeutic methods.[5–9] Separation membranes consisting of composites with 2D materials as a major component are also an emerging application with the potential to outperform the state of the art technology.[10] 2D nanomaterials can be created and used as colloidal dispersions or supported on substrates,[11] and this versatility makes them very exciting to the wider scientific community, as well as reinvigorating interest in the bulk materials from which they are derived from such as graphitic carbon, black phosphorus and molybdenite (MoS_2).

[a]*School of Chemistry, The University of Manchester, Oxford Road, Manchester M13 9PL, UK. E-mail: aleksander.tedstone@postgrad.manchester.ac.uk*
[b]*Laboratory of Inorganic Chemistry, University of Helsinki, P.O. Box 55 (A.I. Virtasen aukio 1), FI-00014 Helsinki, Finland. E-mail: john.brent@helsinki.fi*
[c]*School of Materials, The University of Manchester, Oxford Road, Manchester M13 9PL, UK. E-mail: david.lewis-4@manchester.ac.uk*

2 Two-dimensional (2D) nanomaterials

Two-dimensional (2D) nanomaterials are ultra-thin materials derived from bulk crystalline solids with intrinsically layered atomic structures. These bulk materials are also known as van der Waals solids by virtue of having no molecular bonds between layers and are held together by non-covalent interactions. The archetypal 2D nanomaterial is graphene, the atomically thin material derived from bulk graphite, the properties of which were elucidated in 2004, and sparked interest in the exploration of analogous materials with different compositions. Since this discovery, various allotropes of other main group elements have been identified as 2D material candidates, as have materials comprised of more than one element that exhibit layered structures. The extremely thin nature of these materials can give rise to unique properties and facilitates the creation of layered heterostructures of different 2D materials, opening up a massive field of potential study that has already had significant achievements and may become the basis of emerging technologies that rely on these unique material properties. Not all 2D materials have known bulk counterparts, although in principal any sufficiently stable monolayer compound should be capable of forming an extended multi-layer crystal.

2D materials can be prepared by a number of routes, principally falling into the categories of top-down or bottom-up preparation.[12] Top-down routes typically involve the naturally occurring parent material being processed into 2D form *via* exfoliation of layers, either mechanically, in solution or *via* intercalation–expansion. Bottom-up routes utilise the direct synthesis of the material in controllable conditions such as chemical vapour deposition (CVD), solution syntheses or gas-phase methods such as physical vapour transport (PVT). Variants of each of these techniques are legion and judicious selection is required for any application as each technique carries its own benefits and disadvantages.

Where a 2D material has a naturally occurring bulk parent material, it is important to note that some properties only arise at the few- and mono-layer length scales due to confinement of electronic and magnetic states in the dimension of thickness reduction. Not all research or application requires monolayer sheets, and few-layer forms of these materials can be considered 2D depending upon the property of interest, but the distinction is important and context dependant.

2.1 Homoatomic main group 2D materials

There are a number of homoatomic main group analogues of graphene, many of which have been isolated or synthesised in monolayer form in recent years, and some of which remain theoretical possibilities not yet empirically proven to be stable. Silicene, germanene, stanene, and phosphorene have been the subject of reviews and comparative studies of graphene analogues in 2014 and 2015,[13–15] and more recent developments will be discussed herein. The topological insulator and photo-induced topological superconductor properties of silicene, germanene and stanene are collectively discussed in the work of Ezawa,[16,17] – and it is notable that the heavier group IV 2D materials have more in common with each other than with graphene in term of structure and properties. No 2D allotropes of lead or bismuth have been proposed as stable compounds at the time of writing.

2.1.1 Graphene. Graphene has been the subject of great interest and investment,[18] and is the archetypal 2D nanomaterial that has spurred on interest in related 2D structures. Potential technological applications are numerous due to its exceptionally high conductivity and tensile strength. The material is chemically stable in atmospheric conditions yet readily functionalized, and has many derivatives, for example graphene oxide, graphane and fluorographene.[19–22] A principle challenge in bringing graphene into real-world applications is the size and homogeneity of the sheets that can be produced – although efforts and improvements have been made in this area and wafer-scale CVD graphene synthesis has been demonstrated as a viable route for electronics-grade material.[23,24] It is also possible to create superstructures of non-covalently linked graphene crystals for certain applications that do not require an extended homogenous sheet.[25] CVD growth on a liquid Cu substrate can allow isotropic growth of the graphene and this minimizes the density of grain boundaries in the material; crystal defects that have a detrimental effect on graphenes desired properties.[26] The review literature on graphene is extensive so will not be covered in depth in the current work, which will instead focus on more recently proposed and isolated 2D materials.

2.1.2 Silicene. A layered allotrope of elemental silicon analogous to graphite is not found in nature, but this has not prevented theoretical and experimental studies into a monolayer silicon material analogous to graphene.[27,28] Silicene was first proposed in 2009[29] and first isolated in the form of nano-ribbons on a Ag(110) substrate the following year.[30] More recent reports generally focus on Ag(111) surfaces as substrates for silicene growth, yielding a metallic or semi-metallic material.[31–36] It adopts a buckled honeycomb structure,[37,38] in contrast to the planar honeycomb structure of graphene. It has also been shown to grow on Ir(111),[39] a zirconium diboride (0001) surface,[40] and most recently, a zirconium carbide (111) surface.[41] The influence of the substrate is critical for the formation of silicene, as interaction between substrate surface atoms and the silicon atoms forming the material at the surface determines the product phase.[42–44] The substrate can also influence the semiconducting behaviour of the material by opening a semiconducting band gap *via* non-covalent substrate–silicene interactions, an approach which preserves the structure and hence carrier mobility of the material.[45] The Dirac-cone structure responsible for many of graphene's interesting properties is present in 'ideal' or free-standing silicene but highly dependent on the support substrate[43] and silicene could theoretically exhibit a new quantum state of matter: the valley-polarized quantum anomalous Hall state (Fig. 1).[46]

Multilayer silicene films exhibit some air stability, with limited oxidation occurring during a 24 hour period, as the outermost layer provides

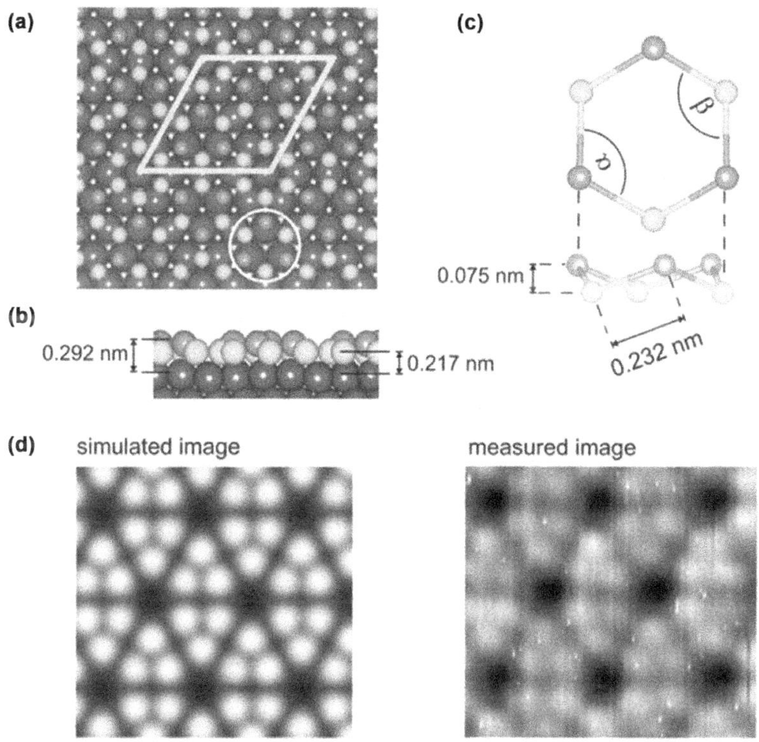

Fig. 1 DFT results for silicene on Ag(111). (a) Plan view of the fully relaxed atomic geometries of the model for silicene on the Ag(111) surface from Fig. 3. (b) Side view of (a). (c) Enlarged image of the hexagonal silicene ring indicated by the white circle in (a). (d) Simulated STM image (left) for the structure shown in (a). The simulated image exhibits the same structural features as those observed in the experimental STM image (right), *i.e.*, a hexagonal arrangement of the triangular structure around dark centers. Reprinted figure with permission from Patrick Vogt *et al.*, Physical Review Letters, 108, 155501, 2012. Copyright 2012 by the American Physical Society.

sacrificial protection from air oxidation, forming silicates.[47] To fabricate devices, strategies must be utilised to prevent degradation in air, which has been successfully achieved in the creation of a silicene field effect transistor which demonstrated the as-predicted Dirac-like character of the material.[48] Oxygen adatoms are capable of inducing a band gap in silicene that can be modulated from semimetallic to semiconducting in nature, and is facilitated by the surface reactivity of silicenes sp^3 hybridized (and therefore reactive compared to graphene's sp^2 hybdrization) surface atoms.[49] Silicene has a lower predicted thermal conductivity than that of graphene, attributed to the differing phonon transport properties of silicenes buckled structure *versus* graphenes planar structure.[50–52] First principles calculations of nitrogen-based gas adsorbates onto the surface of silicene show that they can induce p-type semiconducting behaviour by charge transfer, and yield a carrier mobility of three orders of magnitude greater than that of room temperature graphene, giving it great potential as a gas sensor.[53] This induced band gap was also found to be tuneable with other small organic absorbent molecules.[54]

Hydrogenation of bilayer silicene to silicane has also been demonstrated as a route towards direct semiconducting properties *via* first principles calculations,[46,55] and could be induced by electric field.[56] The hydrogenation of silicene can be reversibly performed and may have potential for reversible property tuning and controlled hydrogen storage.[57]

Transition metal decoration of silicene has been proposed with cobalt being the most promising candidate for stabilization of the quantum anomalous hall state.[58] Various heavy metal adsorbants can induce p- and n-type doping and modification of the intrinsic semiconducting nature in silicene monolayers as potential materials for tunnelling field effect transistors (TFETs) towards low energy electronics.[47] Additionally, external fields can modulate the properties of silicene, for example inducing ferromagnetism has potential for applications in spin-valleytronics.[59,60] Silicene offers great potential for modification of its electronic and magnetic properties *via* atomic doping.[61]

Truly free-standing silicene remains a synthetic challenge, but routes to supported silicene that preserve its theoretically predicted Dirac-cone have been developed, for example those using other 2D materials with minimal substrate-silicene interaction.[62,63] Other non-metallic substrates (CaF_2 and Cl-passivated Si) that minimise the bandgap opening have also been proposed;[64] non-metallic substrates reduce the covalency interaction of Si p_z bands with the substrate to yield only a van der Waals interaction theoretically present between layers of free-standing silicene. Intercalation of alkali metals between silicene and metallic substrates offers another route towards the recovery of silicenes Dirac-cone.[65] The mechanical properties of silicene have been determined by computational methods but not yet empirically verified due to the difficulty in obtaining free-standing material.[66]

2.1.3 Germanene. The germanium analogue of silicene was first predicted in 2009, and germanane – the hydrogenated derivative of germanene – was experimentally isolated in 2013. Germanene was subsequently isolated in 2014 *via* synthesis on a Pt(111) substrate,[67] and subsequently on Au(111) and Al(111) surfaces, adopting several phases, including an almost atomically flat honeycomb structure.[68,69] Other stable phases have been predicted by DFT calculations, including a dumbbell-like honeycomb structure.[70,71] Multilayer germanene has been synthesised on a silicon substrate from an SiGe thin film, providing a potential route for large scale synthesis necessary in electronic applications (Fig. 2).[72]

The propensity of germanene to structurally buckle leads to enhanced chemical reactivity at its surface relative to graphene and susceptibility to hydrogenation – producing germanane.[73] Hydrogenation and halogenation of germanene offer methods of converting it into a topological insulator, with sizeable bandgaps when the material is subjected to strain that could allow practical use, as current 2D topological insulators are hard to integrate into electronic applications owing to their small gap sizes.[74] Surface-functionalised germanene can be further modified *via*

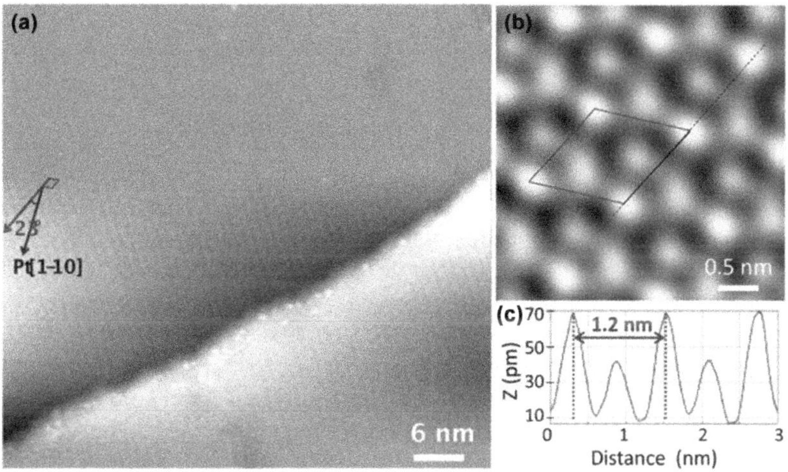

Fig. 2 (a) Occupied-state STM image (potential: $U = -1.45$ V, current: $I = 0.25$ nA), showing a ($\sqrt{19} \times \sqrt{19}$) superstructure of the germanium adlayer formed on the Pt(111) surface. The direction of this reconstruction is indicated by the left arrow. The close-packed direction Pt[110] is indicated by the right arrow. The angle between the arrows is about 23°. (b) Zoomed-in STM image ($U = 1$ V, $I = 0.05$ nA) of the germanium adlayer. (c) Line profile along the dashed line in (b), revealing the periodicity of the germanium superstructure (1.2 nm). Reprinted with permission from L. Li, S.-z. Lu, J. Pan, Z. Qin, Y.-q. Wang, Y. Wang, G.-y. Cao, S. Du and H.-J. Gao, *Advanced Materials*, 2014, **26**, 4820 © 2014 WILEY-VCH Verlag GmbH & Co. KGaA, Weinheim.

strain engineering for further band gap control.[75] The hydroxylization of the surface is also predicted to yield useful modulations in electronic properties by inducing half-metallicity or band gap-opening dependent on the extent of functionalisation.[76]

Germanene's electronic and magnetic properties can be strongly influenced by adsorption of various metal adatoms, which are capable of inducing semiconducting behaviour, as well as metallic and half metallic behaviour with adsorbed 3d transition metal atoms.[38] There is a consensus that alkali metals in particular can be strongly adsorbed to the germanene surface and transfer charge across the 2D material to induce n-type semiconducting behaviour.[77,78]

Germanene nanoribbons edge states are predicted to be strongly nonlinear and therefore unable to exhibit the gapless Dirac behaviour of similar 2D nanoriboons that is desired in various applications,[79] and most types of defect in the layer also prevent Dirac behaviour.[80] As with silicene, it is a challenge to produce free-standing or quasi-free-standing germanene on a substrate which does not destroy its desirable Dirac-cone behaviour.[81] The materials GaS, GaSe, GaTe and InSe offer potential substrates for supporting germanene whilst preserving its desirable neutrality and charge carriers with low effective masses, with a bandgap of 0.14–0.16 eV.[82] First principles calculations indicate that another route towards the preservation of the Dirac-cone in germanene could be to create germanene/germanane heterostructures.[83]

It is predicted the germanene is capable of physi- and chemi-sorbing a number of small gas molecules, and is capable of dissociating O_2 into two

O atoms at its surface.[84] This has potential for the functionalisation of germanene – for example NO_2 induces strong hole-doping of the material when chemisorbed onto it.

2.1.4 Stanene. Stanene, a two-dimensional analogue of tin has received interest stemming from the discovery of other group IV materials, graphene and silicene.[85] As with graphene, stanene displays a number of exciting properties with great potential in electronics. Free-standing Sn nanosheets are not planar, but display a similar buckled structure to silicene and phosphorene, due to a preference of Sn for sp^3 hybridisation and as a result of relatively weak π-bonds.[86] The greater mass of tin atoms compared to carbon and the effect of the material's slightly buckled geometry serves to enhance spin-orbit coupling (SOC) and produce a non-zero bandgap material.[85] These properties are essential to the evolution of the quantum-spin Hall (QSH) effect, which is promising for applications in spintronics and quantum computing.[87] At present the QSH effect can only be achieved at low temperature, therefore topological insulators (TIs) such as stanene, silicene and germanene,[17,88] are sought in order to increase the operating temperature. TIs behave as an insulator but also contain highly conducting edge states.[89] First-principles calculations have also shown that in nanoribbons of the hydrogenated form of stanene, called stanane, which has gapless edge states which display Fermi velocities comparable to that of graphene, while maintaining a non-negligible band gap at the Γ point of the Brilloiun Zone.[87] Silicene[90] and germanene[91] are also promising candidates for the manifestation of the QSH effect. The finite band gap and nature of the edge states may be manipulated by in-plane and out-of-plane magnetic fields and interaction with superconducting materials,[92] chemical functionalisation[87] and strain.[93] When applying an out-of-plane field, at the critical value of 0.5 $VÅ^{-1}$ or under extreme quantum confinement, the material reverts to insulating behaviour.[94]

Topological superconductivity, another potential route to quantum computing, may itself emerge as a result of interaction of the honeycomb lattice with photo-irradiation.[16] Further to electronic applications, stanene has shown promise as a trap for several common pollutants and is expected to be particularly efficient at adsorbing sulfur- and nitric oxides.[95] The strong interaction of the Sn surface with CO and CO_2 leads to considerable weakening of the C–O bonds, indicating potential applications in methanation catalysis.[95]

In the bulk, semiconducting α-Sn undergoes a structural transformation at 286 K to produce β-Sn, which displays metallic behaviour.[96] Three potential morphologies of 2D tin have been proposed, high-buckled- (HB),[97] low-buckled- (LB)[98] and dumbbell (DB)[87]-stanene, though only the low-buckled phase has been successfully synthesised.[99] Synthetic procedures remain rare, therefore the majority of available data is reported *via in silico* experiments. It is the LB form which has therefore been most fully investigated and which is potentially the most promising for the generation of electrical conduction with zero dissipation at room temperature. Early models show that 2D tin displays similar unit cell parameters to the sp^3 diamond-like Sn structure and is stable up to 700 K.[93] For comparison, pristine silicene remains stable up to around 1500 K.[100]

The synthesis of stanene was carried out at room temperature *via* molecular beam epitaxy (MBE) on a Be_2Te_3 substrate. It was found that the film grows along the (111) direction of α-Sn, and the film displays a step-height of ∼0.35 nm, consistent with the expected Sn (111) interlayer distance. No formal bonds are formed between the film and the substrate, rather Sn is most likely physisorbed at the Bi_2Te_3 face-centred cubic (FCC) site. Despite interacting only weakly with the Be_2Te_3, the substrate imparts a small amount of compressive strain on the stanene lattice which ensures that metallic behaviour is retained in this system. These results imply that further functionalisation of the stanene film, or deposition on a substrate which imparts less compressive strain is required to promote the finite band gap necessary for access to QSH effects. It is understood that the majority of the contribution to electronic states around the Fermi level is from the Sn p_z orbitals forming π and π* orbitals.[93] In pristine stanene these remain unsaturated and are therefore easily influenced by the environment, especially by adsorbates. Therefore, there exists significant potential for chemical functionalisation and tuning of electronic states *via* chemical methods.[99] However, it has been calculated that in stanene nanoribbons (SNRs), the density of states (DOS) around the Fermi energy is dominated by valence $p_{x,y}$ orbitals, which remain relatively unaffected by the extent of sp^2/sp^3 hybridisation.[101] Increasing width in SNRs monotonically decreases quantum confinement and therefore increases the band gap; as width tends to infinity dirac cones are recovered. Engineering SNRs to have zigzag-(ZZ) terminated edges also imparts magnetic edge structure, allowing further tuning of the material's electronic and magnetic behaviour.[101]

The Au(111) surface has also been proposed as a potential substrate for stanene deposition.[86] First principles calculations indicate that Sn binds strongly to the gold FCC unit cell, a process driven by significant charge transfer from Sn to Au and the formation of formal covalent bonds. Charge transfer from Sn enhances the extent of sp^2 hybridisation and therefore results in a more planar Sn structure. It is predicted that the enhanced π-overlap, in conjunction with SOC and depletion of electron density in the p_z orbitals as a result of charge transfer, induces a small band gap and therefore Sn's properties as a topological insulator to be retained.[86] Other reports have suggested that the Ge(111) surface is ideal for stabilisation of Sn nanosheets while maintaining α-Sn structure and a non-zero band gap.[98] BN,[102] SrTe, PbTe, BaSe and BaTe have also been proposed.[103]

Electrical properties of stanene nanosheets are significantly affected by their strain environment.[104] Compressive strain increases the energy of the overlapping s and $p_{x,y}$ orbitals and decreases the band gap whereas tensile strain produces the reverse result.[93] Substrate interactions also contribute, depending on the extent and direction of charge transfer between the two; for example, QSH effects can be induced in stanene on

BN under 6% tensile strain whereas Sn nanosheets on AlN substrates require 10% strain.[104] Uniaxial strain has the added effect of breaking sublattice symmetry in inducing a larger energy gap.[93] It has been calculated that by combining strain and an external electric field, the band gap of Sn nanoribbons can be increased by ~120% relative to its intrinsic state.[93]

A form of stanene with dumbbell units, rather than the buckled arrangement has been proposed.[88] The dumbbell arrangement occurs due to the instability of LB stanene, arising from relatively weak π–π bonding. The result produces variable coordination of Sn within the lattice, with 40% threefold-coordinated, out-of-plane and 60% fourfold coordinated, in-plane atoms.[87] Those atoms with unsaturated coordination are likely to be chemically reactive, allowing further modification of the electrical properties.[87]

The hydrogenated form of the material, dumbbell stanane, is expected to display a topological band gap of up to 312 meV and therefore properties such as the QSH effect.[88] Further enhancement of these properties and others such as spin-polarisation may be possible through doping with magnetic elements.[105] In freestanding sheets DB stanene represents a metastable local minimum energy, which may be easily disrupted by the adsorption of group (IV) adatoms, however certain substrates can significantly stabilise the sheets.[71] h-BN and InSb (111) have been proposed as potential substrates for the stabilisation of DB stanene, due to the small lattice mismatch.[88] On boron nitride, the properties of freestanding DB stanene sheets are preserved, InSb induces changes to the electronic behaviour though it may be possible to retain properties such as quantum Hall effects and topological superconductivity using standard semiconductor manipulations such as doping and gating.[88]

The effects of functionalisation on the properties of stanene have recently been probed *via* computational modelling.[106,107] As previously discussed, in the simplest case this often takes the form of hydrogenation of the sheet, especially in order to saturate dangling bonds in DB stanene and form $Sn_{10}H_4$ stanane structures.[87] Others have investigated the use of functionalisation in order to control pore size and therefore gas permeability, finding that hydrogenation reduces pore size and permittivity due to steric effects. Fluorination increases pore size by flattening the sheet and can greatly enhance selectivity for helium.[106] Fluorination also allows band gap widening in LB stanene to up to 300% of the intrinsic value. Functionalisation with CN groups also allows band gap adjustment and maintains stanene's topological properties.[107] Similarly, the coordination of ethyne to form SnC_2H films which interact only weakly with a boron nitride substrate has been shown to retain a large band gap and topological insulator behaviour.[102] Defect engineering as a route to manipulating the properties of graphene and MoS_2 have been well explored.[108,109] In stanene, four potential grain boundary structures have been proposed, which behave as one-dimensional semiconducting or metallic depending on their geometry.[110]

Stanene has received considerable research attention through computational studies due to its exciting potential for quantum computing and spintronics.[102,107,111] Synthetic procedures remain scarce despite great interest; however detailed reports on a wide range of potential substrates may lead to further developments in the field.[86,102,103] Further investigation of 2D Sn on these substrates may also yield experimental observations of the QSH effect, which has not yet been observed in stanene films due to strongly interacting substrates.[99]

2.1.5 Phosphorene. Black phosphorus (BP) is the most thermodynamically stable allotrope of elemental phosphorus although does not exist widely in nature. Its 2-dimensional analogue 'phosphorene' has received considerable attention since its isolation in 2014 due to its intrinsic band gap and high carrier mobility.[49] BP exists in discreet, AB stacked, sheets of sp^3 hybridised phosphorus atoms with an interlayer distance of $\sim$5.2 Å.[112,113] Each atom is bonded to two in-plane neighbours at an angle of 99° from one another and a third, out-of-plane at 103°.[113,114] In contrast to graphene, the sheets are not planar, but 'buckled' or 'puckered', due to phosphorus' lone pair of electrons, which remain relatively localised, rather than contributing to an extended π-system. This structural anisotropy produces significant differences between the properties of the armchair (AC) and zigzag (ZZ) crystallographic directions.

As with all layered materials, exfoliating black phosphorus to a few atomic layers significantly alters its physical and chemical properties.[49,115–118] Most notably, it results in bandgap widening (from $\sim$0.3 eV in bulk BP to 1.5–2 eV in the monolayer) and exceptionally high hole mobility up to 1000 $cm^2\,V^{-1}\,s^{-1}$.[49,117,119] These features have already led to the investigation of phosphorene and few-layer black phosphorus (FLBP) for use as the active material in FETs and CMOS logic devices,[49] gas sensors[120] and photodetectors,[121] thermoelectric devices[122] and energy storage.[123] Moreover, the semiconducting nanosheets are complimentary to graphene and may facilitate the fabrication of flexible electronic devices on an industrial scale.[124,125]

The influence of thickness of phosphorene sheets on their behaviour has been studied both *in silico* and using micromechanically exfoliated samples. The findings are of particular importance for the use of BP as the active layer in FETs and solar energy harvesting devices. As noted above, thinning the material to a few atomic layers is accompanied by a monotonic increase in the electronic band gap[126] and also increases the work function and height of Schottky barriers.[127] The effect of thickness on field effect mobility of charge carriers has been assessed by at least two groups,[117,128] both finding a reducing in mobility compared to bulk values, though with differing optimum thicknesses for maximum mobility of 5 and 10 layers. It is clear that FLBP outperforms monolayer devices in terms of effective mass of carriers,[127] as well as being intrinsically more chemically stable.[49] In experimental devices, measured mobility has fallen short of the potential values predicted from DFT studies, an effect attributed to non-passivated surfaces and chemisorbed species, difficulty in aligning contacts along crystallographic axes and contact resistance at metal/phosphorene junctions.[49] Device ON/OFF ratio is also dependant on flake thickness, with maximum values of 10^5 observed in 2 nm thick devices. Device performance is improved *via* the use of high κ dielectrics such as alumina[129] and HfO_2.[130,131] Top-gated devices are common since this provides the additional benefit of encapsulation of the active material, thus preventing adsorption of gaseous species and/or degradation of the phosphorene surface.[132,133] The combination of phosphorene tranistors with graphene contacts and a h-BN capping layer has led to the formation of fully 2D device architectures.[134] If BN is also used as a buffer material between individual phosphorene nanosheets, it is possible to maintain the characteristics of monolayer phosphorene in restacked BP films.[135]

Due to its recent discovery, there remain relatively few reports discussing the mechanical and electronic characteristics of phosphorene flakes. Several groups have modelled the material's behaviour under a variety of circumstances using density functional theory, with particular attention paid to the influence of strain and edge-shape on electronic properties.

Density functional theory predicts that the geometry of the edge sites in 2D BP has a significant impact on its electronic and mechanical properties; this is especially true when considering phosphorene nanoribbons (PNRs), which have a greater proportion of edge sites relative to nanosheets. The presence of edge sites induces gap states in the phosphorene band structure.[136] Zigzag sites are metallic under most circumstances,[136,137] whereas armchair sites may be insulating[136] or semiconducting, depending on ribbon width.[137] Zigzag-terminated phosphorene nanoribbons (Z-PNRs) are more susceptible to quantum size effects and undergo bandgap widening from 1.4 to 2.6 eV when ribbon width is reduced from 26 Å to 6 Å.[138] Passivation of edge sites with hydrogen restores semiconducting characteristics[46,136] but may lead to the formation of indirect-gap material in Z-PNRs.

Adjustment to electronic states may be induced by imparting relatively small levels of strain. Compressive or uniaxial strain produces bandgap narrowing in monolayer and bilayer phosphorene. The bandgap is more sensitive to compression,[139] though may be modulated continuously under uniaxial strain to induce a direct–indirect bandgap transition[139] and eventually metallic characteristics[140] in monolayers. Controlled use of strain allows enhancement of phosphorene's existing orientation-dependant conductivity, which arises from its structural anisotropy.[141] In its native form phosphorene is a promising thermoelectric material due to perpendicular preferred directions of electronic and thermal conductivity. Electrical conductivity is approximately one order of magnitude higher along the armchair direction[142] whereas thermal conductivity is higher in the zigzag direction.[122] This effect may be enhanced *via* the application of 5% strain, which causes band convergence, greatly improving electronic conductivity and the performance of black phosphorus as a thermoelectric material.[128] Higher levels of strain may considerably alter the atomic arrangement within black phosphorus sheets. Dong and coworkers have calculated that monolayer phosphorene loses its characteristic puckered structure at under 12% isotropic tensile strain, adopting a flat, hexagonal structure more similar to graphene. In bilayer samples the interlayer distance contracts and approaches the P–P bond length, implying that the layers can become formally bonded at 22% strain. It is unclear whether subsequently relaxing the system would cleave the P–P interlayer bonds or form a (meta)stable non-layered phosphorus allotrope.

Bulk BP was initially synthesised in 1914 from white phosphorus under high pressure (12 000 atm) and moderate temperature (200 °C).[143] In the original work Bridgman acknowledged the presence of a small amount of amorphous black phosphorus within the microcrystalline product. Jacobs later confirmed that both higher temperature and pressure are required to form exclusively crystalline material and partially elucidated the mechanism of transformation between white and black phosphorus.[144] Black phosphorus was noted for its high density and stability relative to other forms of phosphorus and for exhibiting good electrical and thermal conductivity.[143] Subsequent work focussed on improving the synthesis of BP, specifically in addressing the requirement for high pressure, by carrying out the synthesis within molten bismuth[145] and mercury,[146] and more recently *via* the use of tin iodide mineralising agents to facilitate the use of low-pressure thermal-transport reactions.[147–149]

Despite its promising characteristics, BP received relatively little attention until very recently. Interest from the wider scientific community was sparked after the fabrication of few-layer BP field effect transistors (FETs)[117] and the isolation of single-layer phosphorene flakes.[49] Both of these reports, which emerged independently in early 2014, used micromechanical exfoliation to obtain ultrathin BP. Such methods are commonly used in the production of pristine sheets research purposes but are inherently unscalable. Lewis and co-workers reported the exfoliation of black phosphorus nanosheets in *N*-methyl-2-pyrrolidone (NMP) to produce FLBP *via* a scalable, liquid-exfoliation method. Since then various groups have reported the exfoliation of BP in various solvents including dimethylformamide (DMF), dimethyl sulfoxide (DMSO),[150] and cyclohexyl-2-pyrrolidone (CHP).[151]

Although black phosphorus is known to be the most stable form of elemental phosphorus, phosphorene itself has been shown to degrade rapidly under ambient conditions.[152] Exposure of freshly cleaved phosphorene or FLBP nanosheets to the atmosphere results in the formation of droplets on the flake surface, disruption of the sheet structure and electronic properties and etching of the upper layers. This process is measureable within 30 minutes of exposure[132] and damage is observable with the naked eye within 48 h.[118,152] When sheets are immersed in water, this process occurs within 1 hour.[153] The observed droplets are thought to consist of a mixture of phosphines and phosphorus acids.[151] The precise mechanism of degradation remains poorly understood, however water, oxygen and visible light are all thought to play a role[153] and oxidation proceeds from sheet edges, leading to size-dependant stability characteristics.[151] Density Functional Theory calculations have indicated that the reaction is likely to proceed *via* the adsorption of O_2 on the BP surface. Interaction with the phosphorus lone pair results in the

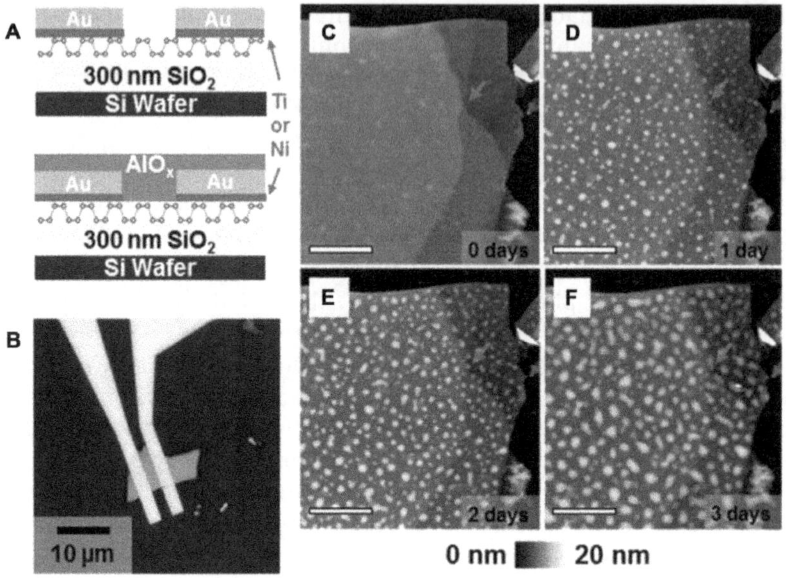

Fig. 3 Time-dependent degradation of exfoliated black phosphorus (BP) flakes on a SiO_2 substrate. (A) Schematics of unencapsulated and AlO_x encapsulated BP field-effect transistors (FETs). (B) Optical image of a typical BP FET on 300 nm SiO_2. AFM height images for a 9.0 nm thick unencapsulated BP flake (C) after exfoliation, (D) 1 day in ambient conditions, (E) 2 days in ambient conditions, and (F) 3 days in ambient conditions. The arrow shows the same region on the flake, and the scale bars are 1 μm. Reprinted with permission from J. D. Wood et al., *Nano Letters*, 2014, **14**, 6964–6970. Copyright 2014 American Chemical Society.

dissociation of O_2 and formation of phosphorene oxide which interacts more strongly with water than does pristine phosphorene (Fig. 3).[154]

These considerations have lead groups to address the stability of phosphorene, primarily through the encapsulation of nanosheets under insulating capping layers. Alumina[133] and h-BN[134] have been shown to effectively protect phosphorene from degradation by providing a barrier to oxygen and moisture ingress. As the final stage of oxidation of FLBP leads to the formation of a P_2O_5 surface, it has been postulated that a subnanometre protective oxide may be formed to protect non-surface layers from damage.[155] If this is found to be correct it may be that full encapsulation of nanosheets is unnecessary.

During liquid exfoliation procedures the dispersing medium, (typically an organic solvent), in addition to its primary function of maintaining the colloidal stability of the nanosheets, also provides a barrier to oxidation. Nanosheets exfoliated in CHP degrade according to an exponential decay function, with a rate proportional to both their surface area per unit mass and the water content of the solvent.[151] Under ideal conditions the decay constant of nanosheets may be improved by at least 2 orders of magnitude over those of micromechanically cleaved sheets immersed in water.[151,153] Due to the strongly coordinating nature of solvents such as NMP and CHP, solvent residue often remains bound to the sheet surface after moderate drying. This has allowed the formation of stable, aqueous phosphorene dispersions by initially exfoliating in a basic NMP solution and washing with water.[156] In this case it appears that the NMP remains bound to the sheet surface, preventing direct interaction between phosphorene and water.

The requirement for initial exfoliation in organic solvents has been recently addressed by Hersam and coworkers.[157] The ability of surfactants to impart colloidal stability onto aqueous nanosheet dispersions has been known for some time.[158] In this case, sodium dodecylsulfate (SDS) is used to provide both colloidal and chemical stability. The behaviour of aqueous sols over the course of several days has not yet been reported therefore it is not possible to directly compare the protection offered by surfactants. The flakes were shown to be on average, thinner than those produced *via* exfoliation in NMP with a significant proportion of monolayer sheets, though average lateral size was also reduced. Studies on the exfoliation of MoS_2 have shown the concentration of sheets to be determined by surfactant concentration,[159] in the above study, using 2% (w/v) SDS the concentration of BP was increased by one order of magnitude over NMP. The ability to exfoliate material in both organic and aqueous environments opens a much wider range of possibilities for the functionalisation and manipulation of phosphorene nanosheets compatible with large-scale solution-processing. Recent work from Lewis and co-workers has found that that in aqueous surfactant solution phosphorene degrades to release orthophosphate (PO_4^{3-}) and that breakdown proceeds to the point at which the remaining phosphorene sheets are potentially entirely encapsulated within surfactant micelles.[160]

2.1.6 Arsenene and antimonene. Arsenic and antimony fall directly below phosphorus in group V and also have elemental layer structures, and as such have therefore been investigated following the discovery of phosphorene. Indeed the properties of $As_{1-x}P_x$ alloys have already been investigated computationally and the compositions required to produce structural changes and direct-indirect band gap crossover have been predicted.[161] The monolayer forms of the materials, arsenene and antimonene are expected to have potential applications in LEDs, solar cells and nanoelectronics.[162–164] Both also have the potential to exhibit the quantum spin Hall (QSH) effect *via* minor tuning of their electronic states to produce gapless edge states and are therefore potentially of use in quantum computing (Fig. 4).[165,166]

In their bulk forms both elements display several allotropes and at room temperature display stable, layered,[167] or pseudo-layered[168] structures similar to phosphorene. It is these non-planar structures that the monolayer analogues, arsenene[167] and antimonene[169] are thought to have been synthesised. Others have identified a second potentially stable arrangement of monolayer arsenic which is analogous to blue phosphorene whereas freestanding sheets of planar arsenene are expected to be unstable.[162]

In both arsenene and antimonene, the confinement of the material to two dimensions results in semi-metal to semiconductor transition and the materials display wide band gaps when compared to phosphorene

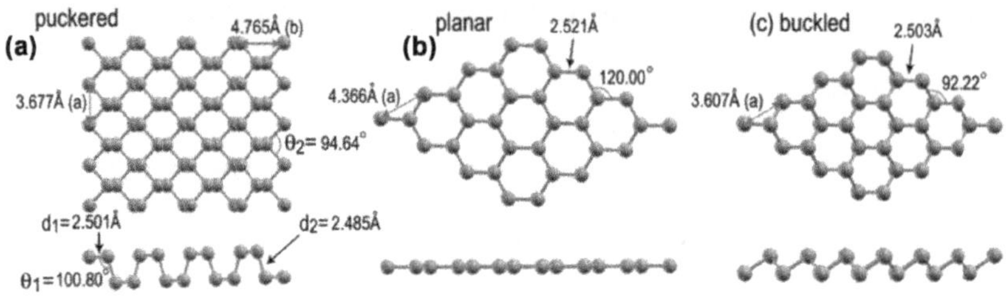

Fig. 4 Optimized geometric structures, phonon dispersion curves, electronic band structure, and density of states of puckered, planar, and buckled arsenenes. Fully optimized structure of (a) puckered (b) planar, and (c) buckled arsenene. The length of the arrow indicates the lattice constant. Reprinted figure with permission from C. Kamal and M. Ezawa, *Physical Review B*, 2015, **91**. Copyright 2015 by the American Physical Society.

and most MX_2 materials.[167] As one would expect, these band gaps are highly influenced by strain,[162,166,170] thickness[170] and heteroatoms.[171] It is possible to induce direct–indirect bandgap transitions in arsenene by substitution of boron and nitrogen into the lattice, while inclusion of oxygen and selenium may induce metallic behaviour.[171] Pristine monolayer arsenic displays an indirect band gap of 1.64 eV, which switches to direct behaviour in the bilayer material.[165,172] The same effect may achieved by imparting 1% strain onto a freestanding arsenene sheet and at 6% compressive strain, metallic behaviour is observed due to increased orbital mixing and overlap of the valence and conduction bands.

True free-standing nanosheets of arsenene have not yet been reported, however recently multilayer 2D nanoribbons of have been successfully produced using a procedure previously used to isolate germanene[72] and violet phosphorene.[169,173,174] Arsenene nanoribbons were synthesised by treating InAs with nitrogen plasma. The nitrogen ions react preferentially with indium to liberate As and form As/InN/InAs stack. The arsenene portion of the film is typically a 14 nm thick polycrystalline As film, which is non-continuous and contains a high concentration of defects. The band gap was estimated to be 2.3 eV, considerably wider than would be expected from relatively thick films. The authors attribute the band gap widening to stacking effects and strong in-plane quantum confinement resulting from the nanoribbon morphology.

Sb/InN/InSb stacks with a $\sim$5 nm antimonene nanoribbon layer have been produced under almost identical conditions.[169] Upon irradiation with orange light the material displays clear photoluminescence signals which increase in intensity at lower temperatures. As with the arsenene preparation, the estimated band gap of 2.03 eV is not thought to be due to the ultrathin nature of the Sb film but is attributed to quantum confinement in the x,y plane and stacking effects.[169] Recently, bilayer antimonene films have been grown on Bi_2Te_3 and Sb_2Te_3 substrates *via* room-temperature deposition of elemental Sb.[175] The antimonene lattice is very slightly distorted due to the $<2\%$ tensile strain imparted by the substrate and the substrates induce partial charge transfer at the Sb/substrate interface as a result of vdW interactions, rather than covalent bonds. The extent of charge transfer has a direct impact on the band structure of the resulting material.[175]

Further work is clearly required to synthesise larger-area and thinner nanosheets of arsenic and antimony. Due to their relatively strong interlayer forces, their isolation by micromechanical cleavage is more difficult.[176] Liquid exfoliation is thought to remain a potential route to the materials,[176] though it will most likely prove more challenging than exfoliation of traditional 2D nanosheets. For these reasons it is perhaps more likely that gas-phase methods such as chemical vapour deposition will prove to be more viable routes to monolayer As and Sb. Some potential substrates for the stabilisation of the growing films and therefore the promotion of epitaxial growth have already been identified.[175,177] Further work is required to identify other potential substrates and investigate their effect on ultrathin films.

2.1.7 **Borophene.**

Elemental boron adopts a wide variety of structures, including a number of crystalline allotropes, as well as polyhedral clusters,[178] nanotubes and fullerene-type species.[179] No naturally occurring layered allotropes exist that could be exfoliated in a top-down approach towards 2D boron layers, although investigations into the possibility of such a material have recently yielded number of possibilities for 2D boron compounds, discussed herein. The numerous potential polymorphs of 2D boron are generally predicted to be metallic in nature, highly anisotropic in terms of structure and properties,[180–182] and are predicted to exhibit superconductivity.[183,184]

Ab initio calculations performed by Wang *et al.* point towards two distinct 2D orthorhombic form of boron that should exhibit massless Dirac fermions, of the space groups *Pmmm* and *Pmmn*, that are predicted to be kinetically stable in ambient conditions.[185] Using an *ab initio* evolutionary structure search, Ma *et al.* propose an ionic form of 2D boron with a *P6/mmm* space group, in contrast to the predominantly covalent structures of other 2D materials such as graphene. Quasi-planar neutral and anionic boron clusters have been experimentally studied, including $[B_{36}]$, $[B_{35}]^-$, $[B_{30}]^-$ and $[B_{40}]^-$, and it is proposed that these could act as building blocks for an extended 2D boron allotrope with periodic holes in its structure.[186–190] Density functional theory calculations indicate that metallization of cluster like this with lithium can afford a hydrogen storage medium.[191] Lithiation may also make 2D boron a promising anode material for Li-ion batteries, with a specific capacity of four times that of a graphite anode material.[192]

Atomically thin boron layers have been experimentally produced on an Ag(111) substrate and demonstrated to be quasiplanar with the as-predicted periodic holes present in the structure.[193] Homogenous and striped phases were found to be present, and susceptible to oxidation in ambient conditions.[194] It is possible that other metal substrates may also be capable of stabilising atomically thin boron layers, such as Ni, Cu, and Au.[195] The corrugated and highly anisotropic material formed differs quite dramatically from other 2D materials and it is thought that free-standing borophene will pose a much greater challenge to isolate than other homoatomic main group 2D materials (Fig. 5).[196]

Boron's electron deficient nature when sp^2 hybridised causes 3 centre, 2 electron bonds to be formed in many boron complexes with empty p_z

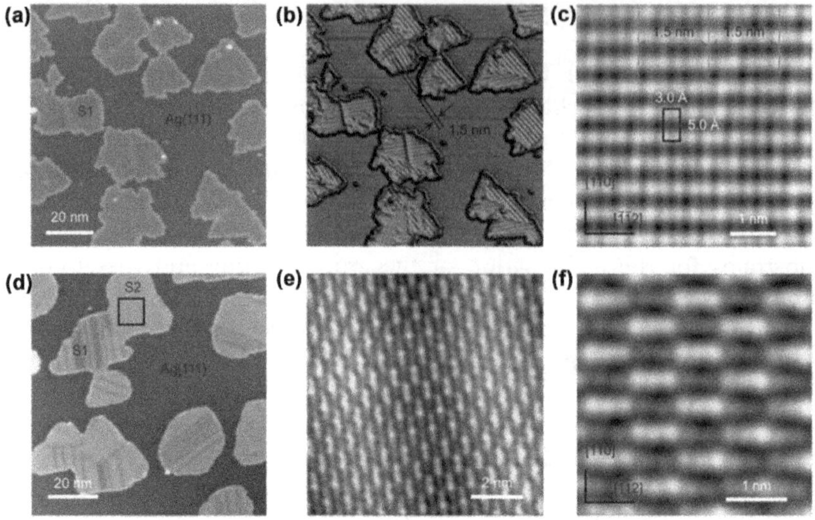

Fig. 5 Formation of 2D boron sheets on Ag(111). (a) STM topographic image of boron structures on Ag(111), with a substrate temperature of ~570 K during growth. The boron islands are labelled as 'S1' phase. (b) Three-dimensional version of (a), in which the stripes with 1.5 nm intervals are clearly resolved. (c) High-resolution STM image of S1 phases. The S1 unit cell is marked by a black rectangle, and the 1.5 nm stripes are indicated by solid lines. (d) STM image of boron sheets after annealing the surface in (a) to 650 K. The two different phases are labelled 'S1' and 'S2'. Most boron islands are transformed to the S2 phase, but the S1 phase still remains in small parts of the islands. (e) STM image obtained on the area marked by the black rectangle in (d). (f) High-resolution STM image of the S2 phase, zoomed from (e). Note that the orientation of the image is rotated to allow comparison with (c). Bias voltages of STM images: −4.0 V (a,b), 0.9 V (c), −4.0 V (d), 1.0 V (e,f). Reprinted by permission from Macmillan Publishers Ltd: Nature Chemistry. B. Feng, *et al.*, 2016, **8**(6), 563–568, copyright 2016.

orbitals extending across the structure, potentially allowing novel band structures.[197] Free standing borophene tends to be unstable relative to its main group 2D counterparts, but should display better stability as hydride terminated sheets, which are predicted to exhibit massless Dirac fermions much like graphene. Similarly, amine functionalisation offers a route towards improving the stability and mechanical properties of 2D boron layers – molecular dynamics simulations indicate that grafting NH_2 groups on boron nanoribbons can increase their structural stability and resistance to high temperature deformation.[198] It is conceivable that borophene may always be reliant on some form of substrate stabilisation or surface functionalisation.

2.2 Heteroatomic main group 2D materials

Combinations of main group elements can also yield van der Waals structures; binary combinations of elements such as boron nitride, tin sulphide/selenide, germanium sulphide/selenide, and lead iodide can form layered compounds.

2.2.1 Hexagonal boron nitride (h-BN). Monolayer hexagonal boron nitride (h-BN) is atomically flat, like graphene, and isostructural – bearing the same relationship to graphene as borazine does to benzene. Bulk boron nitride (sometimes known as 'white graphite') is produced synthetically, and the hexagonal allotrope is easily accessible for large scale production, indeed it is already mass produced for a variety of applications ranging from lubrication to cosmetics. Exfoliation *via* micro-mechanical and liquid phase routes have been demonstrated as a route to few- and mono-layer h-BN and its chemical stability makes this facile under a wide variety of conditions.[199,200] It is also possible to synthesise monolayer h-BN *via* bottom-up fabrication routes; CVD has been demonstrated as a method of making large area high quality h-BN monolayers with controllable morphology and nitrogen/boron ratio.[200–202]

As an extended solid, h-BN exhibits electromagnetic metamaterial properties, specifically that of a hyperbolic metamaterial,[203] wherein the highly anisotropic electronic or magnetic properties of the material can make it exhibit features including strong enhancement of spontaneous emission, diverging density of states and negative refraction, which are rarely observed in nature or conventional materials.[204] Controlling the number of atomic layers of h-BN down to few- and monolayer thickness has been demonstrated as a way of controlling the wavelength of the surface plasmon polaritons exhibited as a consequence of its status as a hyperbolic metamaterial – an observation that could be extended to other 2D nanomaterials and one of fundamental importance in physics.[205]

2D h-BN can exhibit stable voids that migrate under electron[206] or ion beam irradiation[207] at elevated temperatures, and may offer a method of inducing controlled porosity in the material for filtration membranes. It has been observed that care must be taken during sample preparation to prevent unwanted formation of defects and voids, which can take various forms in the monolayer material.[208]

2.2.2 SnS. The group IV–VI family of semiconductor materials includes (among others) SnS, SnS_2, GeS and Sn_2S_3. Several of these display layered structures similar to more well-known nanosheet materials, simplifying the formation of 2D nanosheets. Specifically, α-SnS and GeS are most commonly found in a distorted NaCl structure analogous to (group V) phosphorene, and SnS_2 displays 6-fold coordination of tin in the layered PbI_2 structure. Such materials are of interest as solar energy materials, due to their impressive absorption coefficients, geological abundance of their constituent elements, and low toxicity. In their bulk forms the materials generally display narrow band gaps in the visible region (1.1 eV for SnS, 1.5 eV for GeS,[209] 0.95 eV for Sn_2S_3 and 2.4 eV for SnS_2).

Orthorhombic herzenbergite (α-SnS) is of particular interest as its band gap closely matches that of silicon; in addition its anisotropic, orthorhombic structure allows access to orientation-dependant properties, as in phosphorene. The performance of current champion SnS solar cells ($<5\%$ PCE) still fall well short of their potential, due to issues surrounding stoichiometry, defects, buffer layers and the formation of Schottky barriers at contacts.[210,211] Band-gap tuning is achieved by adjusting particle size[212,213] and stoichiometry.[214,215] Thinning to a few atomic layers also enhances the material's intrinsic anisotropy.

SnS_2 on the other hand exists in layers of hexagonally close packed sulfur with octahedrally coordinated tin ions, the material displays at least 70 polytypes, with variation in the length of the crystallographic *c* axis and therefore the interlayer distance.[216] *Ab initio* investigations indicate that the interlayer distance and band gap are also determined by the stacking order of SnS_2 layers in a bilayer crystal.[217] 5 stacking arrangements were investigated and found to induce a band gap shift of ~0.3 eV and increase the interlayer distance by >0.5 Å.[217]

Density functional theory indicates the combination of SnS and SnS_2 monolayers (SnS@SnS_2) produces heterojunctions with strong electronic interactions between layers.[218] In such a case, the resulting material remains an indirect semiconductor, however additional occupied and unoccupied electronic states arise, with the effect of reducing the band gap to ~1 eV and affecting charge transfer from the SnS_2 to the SnS layer. Doping of both SnS and SnS_2 with selenium has been allows continuous adjustment of their respective band gaps, improving their potential as solar energy materials.[219,220] Through fine adjustment of the S/Se ratio it is possible to shift the band gap of SnX_2 nanosheets by around 1 eV.[220]

Unusually for a layered material, the early studies of SnS nanosheets focussed predominantly on bottom-up rather than top-down procedures. The first top-down fabrication of ultrathin SnS was carried out by liquid exfoliation,[212] rather than the micromechanical route more commonly used in initial investigation of a nanosheet material.[221] Lewis, Seifert and co-workers were able to produce nanosheets with a mean length of around 100 nm by ultrasonicating bulk SnS in NMP. By centrifuging the dispersions further it was possible to isolate highly monodisperse bilayer particles which showed significant band gap widening, which matched well with computational predictions (~0.6 eV). Monolayer nanosheets were typically not observed, indicating a significant energetic barrier to exfoliating beyond bilayers.

Until recently, the majority of research has focussed on the formation of SnS thin films.[215,216,222] Generally such routes produce films of vertically-aligned sheets, rather than epitaxial growth, though substrate patterning, similar to a procedure developed for MoS_2 deposition[223] has produced ultrathin SnS films.[224]

The growth of SnS nanoparticles has received some attention, notably from Eychmüller and coworkers.[213] The synthesis of 2D sheets is less developed, and in several cases, proceeds *via* the orientated self-assembly or ripening of growing nanoparticles.[225–227] A solvothermal route has been developed to produce vertically aligned nanosheet films on a tin substrate.[228] By tuning reaction conditions, sheet thickness as low as 10 nm has been achieved. The first synthesis of colloidal, highly crystalline SnS nanosheets was reported in 2005,[229] and in 2011, Wang and coworkers reported the formation of 20 nm thick sheets with edge lengths of up to 7 μm.[230]

Direct synthesis of SnS nanoplates (*i.e.* preferential sheet growth rather than orientated attachment) from $Sn(acac)_2$ and elemental sulfur has been demonstrated[231] in a low temperature synthesis exploiting the lower energetic barriers to particle growth within the *x*,*y* plane. As well as reaction temperature, the concentration of capping agents and ratio of precursors also determines reaction rate and therefore the final morphology of the product in such reactions.[232]

The formation of nanosheets *via* assembly of growing nanoparticles remains a common route to ultrathin SnS in the liquid phase.[233,234] The formation of size-tunable nanoparticles from a tetrachlorotin–oleate complex and their aggregation allows the formation of 4–15 nm thick platelets.[233] As the platelets are made up of a 'monolayer' of seed nuclei; the thickness of the seeds directly determines that of the product. Orientated attachment is often driven in order to reduce the thermodynamic penalty of exposed high energy faces. Alternatively, under energetic solvothermal conditions, high pressure may reduce the overall entropy of a system and force growing nuclei to coalesce in an ordered manner. These considerations have provided a route to large area (30×3 μm) and exceptionally thin (7 nm) SnS sheets.[235] Films produced from the above sheets showed great promise as NO_2 sensors.

SnS_2 nanosheets are produced in a similar manner, either through attachment of growing seeds[234] or by inhibiting growth in specific crystal facets by coordination of solvents or capping ligands.[236] Uncontrolled aggregation of SnS_2 produced from L-glutatione (L-GSH) and tin chloride has produced 3D microarchitectures made up of sheets 8 nm thick and several 100 nm in length.[237] The enhanced photocatalytic properties of the product are attributed to the high surface are of the nanosheet material.[237] Orientated attachment has allowed the formation of 3–4 nm thick SnS_2 nanosheets, which show enhanced sodium ion storage properties over other SnS_2 and SnO_2 alternatives.[234] The improved performance is again attributed directly to the ultrathin nature of the material and orientated attachment proceeds in order to minimise the exposure of higher-energy faces. Incomplete attachment of nuclei has allowed the formation of highly porous and weakly ferromagnetic SnS_2 nanosheets. In this case magnetism arises as a result of the high proportion of grain boundaries and defects within the highly polycrystalline sheet structure.[238]

SnS_2 has been extensively investigated as a host material for alkali metal ion storage. The inclusion of graphene to form a nanocomposite greatly enhances the conductivity and mechanical stability of the material.[239] This effect has been observed when combining SnS_2 both with 2D graphene nanosheets and 3D graphene foams.[240] The formation of SnS_2/graphene nanosheet composites by growing tin disulfide on a graphene oxide scaffold has the additional effect of reducing the SnS_2 work function considerably and thereby improving the material's field emission properties.[241]

Tin sulfides have potential in a diverse range of applications, notably in solar energy and thermoelectric devices.[215,222,242] Thin film SnS and SnS_2 have been well studied for several decades, however nanoparticles and nanosheets have been much less extensively investigated. The examination of pristine nanosheets obtained from top-down methods will surely yield important data on the fundamental properties of 2D tin sulfide and allow rational device design and optimisation. Several bottom-up methods have been developed, however these rarely obtain highly crystalline and atomically thin nanocrystals; the development of scalable bottom-up routes to thinner nanosheets remains an ongoing challenge.

2.3 Transition Metal Dichalcogenides (TMDCs)

Transition metal dichalcogenides (TMDCs), particularly molybdenum disulfide (MoS_2), have been the focus of a large proportion of the 2D material research after graphene, and as they often possess a substantial band gap, they are complementary in electronics applications to graphene, as well as being robust to oxidation and hydrolysis. Efforts have been made to further modify the parent materials by doping with other atoms, such as transition metals,[243] lanthanides,[244] main group elements, and alloying to produce mixed chalcogenides (*e.g.* MoS_xSe_{2-x}).[245,246] The main candidates for research and application are the chalcogenides of tungsten and molybdenum (MX_2 where M = Mo, W and X = S, Se, Te), as they are particularly well suited to optoelectronic and digital electronic applications, due to their bandgaps being within the visible spectrum of light, and the relatively high mobility of charge carriers generated in the material.[247] A large number of other metal chalcogenides can exist as 2D structures; Ti, Zr, Nb, Ta, Re, and Sn all form layered dichalcogenides, and M_2X_3 phases (M = Bi, Sb and X = S, Se, Te) have also been the subject of recent research efforts.[248–250]

The liquid phase exfoliation of bulk TMDCs is a popular route towards isolation of nanosheets, and can be achieved by sonication,[158,251–254] application of shear force[159,255] or freeze/thaw cycling.[256] Liquid suspensions can then be processed according to the desired use, and have excellent potential for fabrication on large scales, although the size dispersions of the products can vary significantly,[257] and to achieve large flakes of material the process is reliant on very high quality parent material. The bottom-up routes towards 2D TMDCs are generally vapour phase deposition processes, which are not as mild as solvent exfoliation but can yield very high quality large monolayer flakes, and facilitate epitaxial growth – an important factor in creating highly conductive interfaces between disparate materials in 2D nanomaterial heterostructures.[258,259] It is also possible to synthesise 2D TMDCs by solution-based synthesis routes such as thermal decomposition of precursors,[260] hot injection,[261–263] and solvothermal or hydrothermal synthesis.[264–266]

The semiconducting properties of TMDCs make them very attractive for a variety of electronic applications,[267,268] particularly when their optical bandgap corresponds to that of visible light, as this combined with high carrier mobility makes them excellent candidates for solar cell absorbers.[247,269–271] A diverse range of fabricated devices have been reported, including photovoltaic devices, thin film transistors, field effect transistors, and diodes.[272–276]

3 Heterostructures

The combination of different 2D nanomaterials into laminar heterostructures represents a method of achieving a huge variety of properties otherwise inaccessible by traditional materials in the field of optoelectronics. Considering only the number of binary heterostructures possible to engineer from the materials mentioned here yields a great number of combinations. When considering the possibility of multilayer heterostructures containing two or more 2D nanomaterials and the tuning of each material by strategies such as atomic doping, the list becomes near infinite, and rational design will be required as the field matures to direct such heterostructure-based devices towards specific applications. Heterostructures can be vertical (*i.e.* layer upon layer of different materials) or lateral/in-plane/horizontal (*i.e.* different 2D materials in edge-to-edge contact).

Fabrication of 2D heterostructures can give access to a whole range of man-made materials,[203] with controllable electronic and optical properties,[83,277] and novel architectures of existing electronic devices.[278–280] The differences between bulk 3D semiconductor heterojunctions and those formed from 2D nanomaterials are significant and can provide new ways of controlling optical and electronic phenomena, as well as miniaturising existent types of device. It has also been demonstrated that tunnelling field effect transistors (TFETs), promising candidates for low-energy electronics, can be fabricated from a vertical heterostructure comprised of $SnSe_2$/WSe_2. The reported configuration exhibits a nearly broken gap configuration, allowing tunnelling to be switched on and off by application of only a small bias across the junction.[281]

2D materials are excellent candidates for catalytic applications as they have high surface area to volume ratios and are suited to photocatalysis,[1,262,282,283] electrocatalysis and heterogeneous catalysis.[284] Their properties can be further tailored by the fabrication of heterostructures for these applications, both electronic and structural, for enhanced catalytic activity.[285–287] Hybrid architectures of lamellar plates attached to nano- or micro-spheres or polyhedra also offer a way of increasing the surface area for catalytic applications, and MoS_2 supported on graphene foam has been demonstrated as an effective hydrogen evolution catalyst,[288–290] as well as a lithium ion storage medium.[291]

Monolayer MoS_2 on graphene exhibits a dramatically decreased bandgap and the edges of these sheets can function as quantum wires with a unique band structure.[292] MoS_2/graphene hybrids and heterostructures have been demonstrated as gas sensors, sodium ion battery electrodes, photodetectors, and catalysts.[278,289,293,294] It has also been demonstrated that epitaxial growth of MoS_2 onto graphene, SnS_2 and TaS_2 can be achieved by chemical vapour deposition,[295] a relatively scalable technique that could be used for device production. 2D materials can be used as intermediaries between other thin layers in device fabrication, for example h-BN has been demonstrated to act as a tunnelling layer between MoS_2 and a metal contact in a transistor to reduce the Schottky barrier height.[296] It is also theoretically possible to modify the Schottky barrier in an arsenene/graphene heterostructure by variation of the interlayer distance, and experimental realisation of this property could allow high levels of control in 2D heterostructures.[297]

4 Conclusions and outlook

A number of two dimensional nanomaterials have been successfully predicted, isolated and studied in the last few years, and represent promising candidates for innovation and understanding in fundamental and applied science. Proposed uses of these materials, particularly those that are now well understood such as graphene and MoS_2, are broad in scope and have the potential to work their way into our everyday lives, as well as enabling research into unique electronic transport properties, electromagnetic metamaterials,[298] and realisation of new states of matter such as the quantum spin Hall effect and valley-polarized quantum anomalous Hall state.[46,299]

The increased diversity of materials beyond graphene is likely to consolidate the position of 2D materials in the field of nanoscience, and it is possible that commercial and industrial technologies will incorporate them into widespread use in the coming years. There are challenges that remain in this field; high-quality, large area monolayers of most of these materials are still inaccessible, although not all applications require this. It is also important to consider whether these materials are more suitable than the current best technologies in terms of performance, cost effectiveness, and the earth abundance and/or toxicity of their constituents. However, the major challenge in this area, especially with the nascent materials described herein, revolve around issues of stability. The materials – especially the main group allotropes – are inherently unstable and at the moment unsuitable for long term usage due to their transience. It is expected that there will be major research in this second phase of the 2D materials 'gold rush' in consolidating what has been discovered thus far and translating this technology into mainstream use.

Notes and references

1 X. An and J. C. Yu, *RSC Adv.*, 2011, **1**, 1426–1434.
2 Y. Jiao, Y. Zheng, M. Jaroniec and S. Z. Qiao, *J. Am. Chem. Soc.*, 2014, **136**, 4394–4403.
3 T.-F. Yeh, C.-Y. Teng, S.-J. Chen and H. Teng, *Adv. Mater.*, 2014, **26**, 3297–3303.
4 N. Zhang, M.-Q. Yang, S. Liu, Y. Sun and Y.-J. Xu, *Chem. Rev.*, 2015, **115**, 10307–10377.
5 K. Yang, L. Hu, X. Ma, S. Ye, L. Cheng, X. Shi, C. Li, Y. Li and Z. Liu, *Adv. Mater.*, 2012, **24**, 1868–1872.
6 W. Yin, L. Yan, J. Yu, G. Tian, L. Zhou, X. Zheng, X. Zhang, Y. Yong, J. Li, Z. Gu and Y. Zhao, *ACS Nano*, 2014, **8**, 6922–6933.
7 L. Feng and Z. Liu, *Nanomedicine*, 2011, **6**, 317–324.
8 C. Chung, Y.-K. Kim, D. Shin, S.-R. Ryoo, B. H. Hong and D.-H. Min, *Acc. Chem. Res.*, 2013, **46**, 2211–2224.
9 D. Bitounis, H. Ali-Boucetta, B. H. Hong, D.-H. Min and K. Kostarelos, *Adv. Mater.*, 2013, **25**, 2258–2268.
10 G. Liu, W. Jin and N. Xu, *Angew. Chem., Int. Ed.*, 2016, **55**, 13384–13397.
11 M. Nasilowski, B. Mahler, E. Lhuillier, S. Ithurria and B. Dubertret, *Chem. Rev.*, 2016, **116**, 10934–10982.
12 P. Samorì, V. Palermo and X. Feng, *Adv. Mater.*, 2016, **28**, 6027–6029.

13 S. Balendhran, S. Walia, H. Nili, S. Sriram and M. Bhaskaran, *Small*, 2015, **11**, 640–652.
14 A. Dimoulas, *Microelectron. Eng.*, 2015, **131**, 68–78.
15 L. Matthes, O. Pulci and F. Bechstedt, *New J. Phys.*, 2014, **16**, 105007–105020.
16 M. Ezawa, *J. Supercond. Novel Magn.*, 2015, **28**, 1249–1253.
17 M. Ezawa, *J. Phys. Soc. Jpn.*, 2015, **84**, 121003–121014.
18 X. Xiao, Y. Li and Z. Liu, *Nat. Mater.*, 2016, **15**, 697–698.
19 T. Kuila, S. Bose, A. K. Mishra, P. Khanra, N. H. Kim and J. H. Lee, *Prog. Mater. Sci.*, 2012, **57**, 1061–1105.
20 Z. Liu, S. P. Lau and F. Yan, *Chem. Soc. Rev.*, 2015, **44**, 5638–5679.
21 V. Georgakilas, M. Otyepka, A. B. Bourlinos, V. Chandra, N. Kim, K. C. Kemp, P. Hobza, R. Zboril and K. S. Kim, *Chem. Rev.*, 2012, **112**, 6156–6214.
22 M. F. Craciun, I. Khrapach, M. D. Barnes and S. Russo, *J. Phys.: Condens. Matter*, 2013, **25**, 423201.
23 J.-H. Lee, E. K. Lee, W.-J. Joo, Y. Jang, B.-S. Kim, J. Y. Lim, S.-H. Choi, S. J. Ahn, J. R. Ahn, M.-H. Park, C.-W. Yang, B. L. Choi, S.-W. Hwang and D. Whang, *Science*, 2014, **344**, 286–289.
24 V. L. Nguyen, D. J. Perello, S. Lee, C. T. Nai, B. G. Shin, J.-G. Kim, H. Y. Park, H. Y. Jeong, J. Zhao, Q. A. Vu, S. H. Lee, K. P. Loh, S.-Y. Jeong and Y. H. Lee, *Adv. Mater.*, 2016, **28**, 8177–8183.
25 M. Zeng, L. Wang, J. Liu, T. Zhang, H. Xue, Y. Xiao, Z. Qin and L. Fu, *J. Am. Chem. Soc.*, 2016, **138**, 7812–7815.
26 M. Zeng, L. Tan, L. Wang, R. G. Mendes, Z. Qin, Y. Huang, T. Zhang, L. Fang, Y. Zhang, S. Yue, M. H. Rümmeli, L. Peng, Z. Liu, S. Chen and L. Fu, *ACS Nano*, 2016, **10**, 7189–7196.
27 H. Oughaddou, H. Enriquez, M. R. Tchalala, H. Yildirim, A. J. Mayne, A. Bendounan, G. Dujardin, M. A. Ali and A. Kara, *Prog. Surf. Sci.*, 2015, **90**, 46–83.
28 L. C. L. Y. Voon and G. G. Guzman-Verri, *MRS Bull.*, 2014, **39**, 366–373.
29 S. Lebegue and O. Eriksson, *Phys. Rev. B: Condens. Matter Mater. Phys.*, 2009, **79**, 4.
30 B. Aufray, A. Kara, S. Vizzini, H. Oughaddou, C. Leandri, B. Ealet and G. Le Lay, *Appl. Phys. Lett.*, 2010, **96**, 3.
31 N. W. Johnson, P. Vogt, A. Resta, P. De Padova, I. Perez, D. Muir, E. Z. Kurmaev, G. Le Lay and A. Moewes, *Adv. Funct. Mater.*, 2014, **24**, 5253–5259.
32 P. Pflugradt, L. Matthes and F. Bechstedt, *Phys. Rev. B: Condens. Matter Mater. Phys.*, 2014, **89**, 205428–205433.
33 G. Prevot, R. Bernard, H. Cruguel and Y. Borensztein, *Appl. Phys. Lett.*, 2014, **105**, 213106–213111.
34 T. Shirai, T. Shirasawa, T. Hirahara, N. Fukui, T. Takahashi and S. Hasegawa, *Phys. Rev. B: Condens. Matter Mater. Phys.*, 2014, **89**, 241403–241408.
35 J. Sone, T. Yamagami, Y. Aoki, K. Nakatsuji and H. Hirayama, *New J. Phys.*, 2014, **16**, 095004–095019.
36 P. Vogt, P. Capiod, M. Berthe, A. Resta, P. De Padova, T. Bruhn, G. Le Lay and B. Grandidier, *Appl. Phys. Lett.*, 2014, **104**, 021602–021607.
37 K. Kawahara, T. Shirasawa, R. Arafune, C. L. Lin, T. Takahashi, M. Kawai and N. Takagi, *Surf. Sci.*, 2014, **623**, 25–28.
38 Z.-L. Liu, M.-X. Wang, J.-P. Xu, J.-F. Ge, G. Le Lay, P. Vogt, D. Qian, C.-L. Gao, C. Liu and J.-F. Jia, *New J. Phys.*, 2014, **16**, 075006–075015.
39 L. Meng, Y. Wang, L. Zhang, S. Du, R. Wu, L. Li, Y. Zhang, G. Li, H. Zhou, W. A. Hofer and H.-J. Gao, *Nano Lett.*, 2013, **13**, 685–690.

40 A. Fleurence, R. Friedlein, T. Ozaki, H. Kawai, Y. Wang and Y. Yamada-Takamura, *Phys. Rev. Lett.*, 2012, **108**, 245501–245506.
41 T. Aizawa, S. Suehara and S. Otani, *J. Phys. Chem. C*, 2014, **118**, 23049–23057.
42 S. Cahangirov, V. O. Ozcelik, L. Xian, J. Avila, S. Cho, M. C. Asensio, S. Ciraci and A. Rubio, *Phys. Rev. B: Condens. Matter Mater. Phys.*, 2014, **90**, 035448–035453.
43 S. K. Mahatha, P. Moras, V. Bellini, P. M. Sheverdyaeva, C. Struzzi, L. Petaccia and C. Carbone, *Phys. Rev. B: Condens. Matter Mater. Phys.*, 2014, **89**, 201416–201421.
44 M. Svec, P. Hapala, M. Ondracek, P. Merino, M. Blanco-Rey, P. Mutombo, M. Vondracek, Y. Polyak, V. Chab, J. A. Martin Gago and P. Jelinek, *Phys. Rev. B: Condens. Matter Mater. Phys.*, 2014, **89**, 201412–201417.
45 N. Gao, J. C. Li and Q. Jiang, *Chem. Phys. Lett.*, 2014, **592**, 222–226.
46 H. Pan, Z. Li, C.-C. Liu, G. Zhu, Z. Qiao and Y. Yao, *Phys. Rev. Lett.*, 2014, **112**, 106802.
47 P. De Padova, C. Ottaviani, C. Quaresima, B. Olivieri, P. Imperatori, E. Salomon, T. Angot, L. Quagliano, C. Romano, A. Vona, M. Muniz-Miranda, A. Generosi, B. Paci and G. Le Lay, *2D Mater.*, 2014, **1**, 021003.
48 L. Tao, E. Cinquanta, D. Chiappe, C. Grazianetti, M. Fanciulli, M. Dubey, A. Molle and D. Akinwande, *Nat. Nanotechnol.*, 2015, **10**, 227–231.
49 Y. Du, J. Zhuang, H. Liu, X. Xu, S. Eilers, K. Wu, P. Cheng, J. Zhao, X. Pi, K. W. See, G. Peleckis, X. Wang and S. X. Dou, *ACS Nano*, 2014, **8**, 10019–10025.
50 X. Gu and R. Yang, *J. Appl. Phys.*, 2015, **117**, 025102.
51 H. Xie, M. Hu and H. Bao, *Appl. Phys. Lett.*, 2014, **104**, 131906.
52 X. Zhang, H. Xie, M. Hu, H. Bao, S. Yue, G. Qin and G. Su, *Phys. Rev. B: Condens. Matter Mater. Phys.*, 2014, **89**, 054310.
53 W. Hu, N. Xi, X. Wu, Z. Li and J. Yang, *Phys. Chem. Chem. Phys.*, 2014, **16**, 6957–6962.
54 T. P. Kaloni, G. Schreckenbach and M. S. Freund, *J. Phys. Chem. C*, 2014, **118**, 23361–23367.
55 B. Huang, H.-X. Deng, H. Lee, M. Yoon, B. G. Sumpter, F. Liu, S. C. Smith and S.-H. Wei, *Phys. Rev. X*, 2014, **4**, 021029.
56 W. Wu, Z. Ao, T. Wang, C. Li and S. Li, *Phys. Chem. Chem. Phys.*, 2014, **16**, 16588–16594.
57 J. Qiu, H. Fu, Y. Xu, A. I. Oreshkin, T. Shao, H. Li, S. Meng, L. Chen and K. Wu, *Phys. Rev. Lett.*, 2015, **114**, 126101.
58 T. P. Kaloni, N. Singh and U. Schwingenschloegl, *Phys. Rev. B: Condens. Matter Mater. Phys.*, 2014, **89**, 035409.
59 B. Soodchomshom, *J. Appl. Phys.*, 2014, **115**, 023706.
60 R.-w. Zhang, C.-w. Zhang, S.-s. Li, W.-x. Ji, P.-j. Wang, F. Li, P. Li, M.-j. Ren and M. Yuan, *Solid State Commun.*, 2014, **191**, 49–53.
61 K. Zberecki, R. Swirkowicz and J. Barnas, *Phys. Rev. B: Condens. Matter Mater. Phys.*, 2014, **89**, 165419.
62 T. P. Kaloni, M. Tahir and U. Schwingenschlögl, *Sci. Rep.*, 2013, **3**, 3192.
63 M. Neek-Amal, A. Sadeghi, G. R. Berdiyorov and F. M. Peeters, *Appl. Phys. Lett.*, 2013, **103**, 261904.
64 S. Kokott, P. Pflugradt, L. Matthes and F. Bechstedt, *J. Phys.: Condens. Matter*, 2014, **26**, 185002.
65 R. Quhe, Y. Yuan, J. Zheng, Y. Wang, Z. Ni, J. Shi, D. Yu, J. Yang and J. Lu, *Sci. Rep.*, 2014, **4**, 5476.
66 R. E. Roman and S. W. Cranford, *Comput. Mater. Sci.*, 2014, **82**, 50–55.
67 L. Li, S.-z. Lu, J. Pan, Z. Qin, Y.-q. Wang, Y. Wang, G.-y. Cao, S. Du and H.-J. Gao, *Adv. Mater.*, 2014, **26**, 4820–4824.

68 M. E. Dávila, L. Xian, S. Cahangirov, A. Rubio and G. L. Lay, *New J. Phys.*, 2014, **16**, 095002.
69 R. Stephan, M. C. Hanf, M. Derivaz, D. Dentel, M. C. Asensio, J. Avila, A. Mehdaoui, P. Sonnet and C. Pirri, *J. Phys. Chem. C*, 2016, **120**, 1580–1585.
70 V. O. Ozcelik, E. Durgun and S. Ciraci, *J. Phys. Chem. Lett.*, 2014, **5**, 2694–2699.
71 V. O. Ozcelik, D. Kecik, E. Durgun and S. Ciraci, *J. Phys. Chem. C*, 2015, **119**, 845–853.
72 H.-S. Tsai, Y.-Z. Chen, H. Medina, T.-Y. Su, T.-S. Chou, Y.-H. Chen, Y.-L. Chueh and J.-H. Liang, *Phys. Chem. Chem. Phys.*, 2015, **17**, 21389–21393.
73 A. Nijamudheen, R. Bhattacharjee, S. Choudhury and A. Datta, *J. Phys. Chem. C*, 2015, **119**, 3802–3809.
74 C. Si, J. Liu, Y. Xu, J. Wu, B.-L. Gu and W. Duan, *Phys. Rev. B: Condens. Matter Mater. Phys.*, 2014, **89**, 10059–10063.
75 P. Liang, Y. Liu, S. Xing, H. Shu and B. Tai, *Solid State Commun.*, 2016, **226**, 19–24.
76 P. Xiao, X.-L. Fan and L.-M. Liu, *Comput. Mater. Sci.*, 2014, **92**, 244–252.
77 Q. Pang, C.-l. Zhang, L. Li, Z.-q. Fu, X.-m. Wei and Y.-l. Song, *Appl. Surf. Sci.*, 2014, **314**, 15–20.
78 J. Yuan, C. Tang, J. Zhong and Y. Mao, *Appl. Surf. Sci.*, 2016, **360**, 707–714.
79 L. Matthes and F. Bechstedt, *Phys. Rev. B: Condens. Matter Mater. Phys.*, 2014, **90**, 105007.
80 J. E. Padilha and R. B. Pontes, *Solid State Commun.*, 2016, **225**, 38–43.
81 L. Zhang, P. Bampoulis, A. van Houselt and H. J. W. Zandvliet, *Appl. Phys. Lett.*, 2015, **107**, 443002.
82 Z. Ni, E. Minamitani, Y. Ando and S. Watanabe, *Phys. Chem. Chem. Phys.*, 2015, **17**, 19039–19044.
83 R.-w. Zhang, C.-w. Zhang, W.-x. Ji, F. Li, M.-j. Ren, P. Li, M. Yuan and P.-j. Wang, *Phys. Chem. Chem. Phys.*, 2015, **17**, 12194–12198.
84 W. Xia, W. Hu, Z. Li and J. Yang, *Phys. Chem. Chem. Phys.*, 2014, **16**, 22495–22498.
85 C.-C. Liu, H. Jiang and Y. Yao, *Phys. Rev. B: Condens. Matter Mater. Phys.*, 2011, **84**, 195430.
86 S. Nigam, S. Gupta, D. Banyai, R. Pandey and C. Majumder, *Phys. Chem. Chem. Phys.*, 2015, **17**, 6705–6712.
87 X. Chen, L. Li and M. Zhao, *Phys. Chem. Chem. Phys.*, 2015, **17**, 16624–16629.
88 P. Tang, P. Chen, W. Cao, H. Huang, S. Cahangirov, L. Xian, Y. Xu, S.-C. Zhang, W. Duan and A. Rubio, *Phys. Rev. B: Condens. Matter Mater. Phys.*, 2014, **90**, 081112.
89 C. L. Kane and E. J. Mele, *Phys. Rev. Lett.*, 2005, **95**, 146802.
90 Z. Tian, Z. Zhao-Yi, C. Yan, C. Xiang-Rong and C. Ling-Cang, *New J. Phys.*, 2016, **18**, 043001.
91 L. Seixas, J. E. Padilha and A. Fazzio, *Phys. Rev. B: Condens. Matter Mater. Phys.*, 2014, **89**, 195403.
92 S. Rachel and M. Ezawa, *Phys. Rev. B: Condens. Matter Mater. Phys.*, 2014, **89**, 195303.
93 B. van den Broek, M. Houssa, E. Scalise, G. Pourtois, V. V. Afanas'ev and A. Stesmans, *2D Mater.*, 2014, **1**, 021004.
94 M. Houssa, B. van den Broek, K. Iordanidou, A. K. A. Lu, G. Pourtois, J.-P. Locquet, V. Afanas'ev and A. Stesmans, *Nano Res.*, 2016, **9**, 774–778.
95 L. Takahashi and K. Takahashi, *Phys. Chem. Chem. Phys.*, 2015, **17**, 21394–21396.

96 G. P. Thomas, M. Paritosh, P. Kjeld, E. C. Niels, M. K. Mads and L. Arne Nylandsted, *J. Phys.: Condens. Matter*, 2009, **21**, 115502.
97 P. Rivero, J.-A. Yan, V. M. García-Suárez, J. Ferrer and S. Barraza-Lopez, *Phys. Rev. B: Condens. Matter Mater. Phys.*, 2014, **90**, 241408.
98 Y. Fang, Z.-Q. Huang, C.-H. Hsu, X. Li, Y. Xu, Y. Zhou, S. Wu, F.-C. Chuang and Z.-Z. Zhu, *Sci. Rep.*, 2015, **5**, 14196.
99 F.-f. Zhu, W.-j. Chen, Y. Xu, C.-l. Gao, D.-d. Guan, C.-h. Liu, D. Qian, S.-C. Zhang and J.-f. Jia, *Nat. Mater.*, 2015, **14**, 1020–1025.
100 T. H. Osborn and A. A. Farajian, *J. Phys. Chem. C*, 2012, **116**, 22916–22920.
101 B. van den Broek, M. Houssa, G. Pourtois, V. V. Afanas'ev and A. Stesmans, *Phys. Status Solidi RRL*, 2014, **8**, 931–934.
102 R.-W. Zhang, C.-W. Zhang, W.-X. Ji, S.-S. Li, S.-J. Hu, S.-S. Yan, P. Li, P.-J. Wang and F. Li, *New J. Phys.*, 2015, **17**, 083036.
103 Y. Xu, P. Tang and S.-C. Zhang, *Phys. Rev. B: Condens. Matter Mater. Phys.*, 2015, **92**, 081112.
104 D. Wang, L. Chen, X. Wang, G. Cui and P. Zhang, *Phys. Chem. Chem. Phys.*, 2015, **17**, 26979–26987.
105 H. Zhang, J. Zhang, B. Zhao, T. Zhou and Z. Yang, *Appl. Phys. Lett.*, 2016, **108**, 083104.
106 G. Gao, Y. Jiao, Y. Jiao, F. Ma, L. Kou and A. Du, *Beilstein J. Nanotechnol.*, 2015, **6**, 2470–2476.
107 W.-x. Ji, C.-w. Zhang, M. Ding, P. Li, F. Li, M.-j. Ren, P.-j. Wang, S.-j. Hu and S.-s. Yan, *Sci. Rep.*, 2015, **5**, 18604.
108 K. Q. Dang and D. E. Spearot, *J. Appl. Phys.*, 2014, **116**, 013508.
109 H. Zhang, G. Lee, C. Gong, L. Colombo and K. Cho, *J. Phys. Chem. C*, 2014, **118**, 2338–2343.
110 Z. Zhu, Q. Sun and Y. Jia, *AIP Adv.*, 2016, **6**, 035012.
111 C. Huang, J. Zhou, H. Wu, K. Deng, P. Jena and E. Kan, *J. Phys. Chem. Lett.*, 2016, **7**, 1919–1924.
112 J. Dai and X. C. Zeng, *RSC Adv.*, 2014, **4**, 48017–48021.
113 A. Brown and S. Rundqvist, *Acta Crystallogr.*, 1965, **19**, 684–685.
114 R. Hultgren, N. S. Gingrich and B. E. Warren, *J. Chem. Phys.*, 1935, **3**, 351–355.
115 S. Zhang, J. Yang, R. Xu, F. Wang, W. Li, M. Ghufran, Y.-W. Zhang, Z. Yu, G. Zhang, Q. Qin and Y. Lu, *ACS Nano*, 2014, **8**, 9590–9596.
116 J. D. Wood, S. A. Wells, D. Jariwala, K.-S. Chen, E. Cho, V. K. Sangwan, X. Liu, L. J. Lauhon, T. J. Marks and M. C. Hersam, *Nano Lett.*, 2014, **14**, 6964–6970.
117 L. Li, Y. Yu, G. J. Ye, Q. Ge, X. Ou, H. Wu, D. Feng, X. H. Chen and Y. Zhang, *Nat. Nanotechnol.*, 2014, **9**, 372–377.
118 L. Kou, C. Chen and S. C. Smith, *J. Phys. Chem. Lett.*, 2015, **6**, 2794–2805.
119 M. Buscema, D. J. Groenendijk, S. I. Blanter, G. A. Steele, H. S. J. van der Zant and A. Castellanos-Gomez, *Nano Lett.*, 2014, **14**, 3347–3352.
120 S. Cui, H. Pu, S. A. Wells, Z. Wen, S. Mao, J. Chang, M. C. Hersam and J. Chen, *Nat. Commun.*, 2015, **6**, 8632.
121 N. Youngblood, C. Chen, S. J. Koester and M. Li, *Nat. Photonics*, 2015, **9**, 247–252.
122 R. Fei, A. Faghaninia, R. Soklaski, J.-A. Yan, C. Lo and L. Yang, *Nano Lett.*, 2014, **14**, 6393–6399.
123 J. Sun, H.-W. Lee, M. Pasta, H. Yuan, G. Zheng, Y. Sun, Y. Li and Y. Cui, *Nat. Nanotechnol.*, 2015, **10**, 980–985.
124 Q. Wei and X. Peng, *Appl. Phys. Lett.*, 2014, **104**, 251915.

125 W. Zhu, M. N. Yogeesh, S. Yang, S. H. Aldave, J.-S. Kim, S. Sonde, L. Tao, N. Lu and D. Akinwande, *Nano Lett.*, 2015, **15**, 1883–1890.
126 S. Das, W. Zhang, M. Demarteau, A. Hoffmann, M. Dubey and A. Roelofs, *Nano Lett.*, 2014, **14**, 5733–5739.
127 Y. Cai, G. Zhang and Y.-W. Zhang, *Sci. Rep.*, 2014, **4**, 6677.
128 Z.-Y. Ong, Y. Cai, G. Zhang and Y.-W. Zhang, *J. Phys. Chem. C*, 2014, **118**, 25272–25277.
129 S. Das, M. Demarteau and A. Roelofs, *ACS Nano*, 2014, **8**, 11730–11738.
130 N. Haratipour, M. C. Robbins and S. J. Koester, *IEEE Electron Device Lett.*, 2015, **36**, 411–413.
131 H. Wang, X. Wang, F. Xia, L. Wang, H. Jiang, Q. Xia, M. L. Chin, M. Dubey and S.-j. Han, *Nano Lett.*, 2014, **14**, 6424–6429.
132 R. A. Doganov, E. C. T. O'Farrell, S. P. Koenig, Y. Yeo, A. Ziletti, A. Carvalho, D. K. Campbell, D. F. Coker, K. Watanabe, T. Taniguchi, A. H. C. Neto and B. Ozyilmaz, *Nat. Commun.*, 2015, **6**, 6647.
133 J.-S. Kim, Y. Liu, W. Zhu, S. Kim, D. Wu, L. Tao, A. Dodabalapur, K. Lai and D. Akinwande, *Sci. Rep.*, 2015, **5**, 8989.
134 A. Avsar, I. J. Vera-Marun, J. Y. Tan, K. Watanabe, T. Taniguchi, A. H. Castro Neto and B. Özyilmaz, *ACS Nano*, 2015, **9**, 4138–4145.
135 G. C. Constantinescu and N. D. M. Hine, *Nano Lett.*, 2016, **16**, 2586–2594.
136 A. Carvalho, A. S. Rodin and A. H. C. Neto, *EPL*, 2014, **108**, 47005.
137 H. Guo, N. Lu, J. Dai, X. Wu and X. C. Zeng, *J. Phys. Chem. C*, 2014, **118**, 14051–14059.
138 X. Han, H. M. Stewart, S. A. Shevlin, C. R. A. Catlow and Z. X. Guo, *Nano Lett.*, 2014, **14**, 4607–4614.
139 T. Hu, Y. Han and J. Dong, *Nanotechnology*, 2014, **25**, 455703.
140 Y. Ding, Y. Wang, L. Shi, Z. Xu and J. Ni, *Phys. Status Solidi RRL*, 2014, **8**, 939–942.
141 H. Y. Lv, W. J. Lu, D. F. Shao and Y. P. Sun, *Phys. Rev. B: Condens. Matter Mater. Phys.*, 2014, **90**, 085433.
142 Z.-Y. Ong, G. Zhang and Y. W. Zhang, *J. Appl. Phys.*, 2014, **116**, 214505.
143 P. W. Bridgman, *J. Am. Chem. Soc.*, 1914, **36**, 1344–1363.
144 R. B. Jacobs, *J. Chem. Phys.*, 1937, **5**, 945–953.
145 Y. Maruyama, S. Suzuki, K. Kobayashi and S. Tanuma, *Physica B + C*, 1981, **105**, 99–102.
146 H. Krebs, H. Weitz and K. H. Worms, *Z. Anorg. Allg. Chem.*, 1955, **280**, 119–133.
147 S. Lange, P. Schmidt and T. Nilges, *Inorg. Chem.*, 2007, **46**, 4028–4035.
148 T. Nilges, M. Kersting and T. Pfeifer, *J. Solid State Chem.*, 2008, **181**, 1707–1711.
149 M. Köpf, N. Eckstein, D. Pfister, C. Grotz, I. Krüger, M. Greiwe, T. Hansen, H. Kohlmann and T. Nilges, *J. Cryst. Growth*, 2014, **405**, 6–10.
150 P. Yasaei, B. Kumar, T. Foroozan, C. Wang, M. Asadi, D. Tuschel, J. E. Indacochea, R. F. Klie and A. Salehi-Khojin, *Adv. Mater.*, 2015, **27**, 1887–1892.
151 D. Hanlon, C. Backes, E. Doherty, C. S. Cucinotta, N. C. Berner, C. Boland, K. Lee, A. Harvey, P. Lynch, Z. Gholamvand, S. Zhang, K. Wang, G. Moynihan, A. Pokle, Q. M. Ramasse, N. McEvoy, W. J. Blau, J. Wang, G. Abellan, F. Hauke, A. Hirsch, S. Sanvito, D. D. O'Regan, G. S. Duesberg, V. Nicolosi and J. N. Coleman, *Nat. Commun.*, 2015, **6**, 8563.
152 O. I. Joshua, A. S. Gary, S. J. v. d. Z. Herre and C.-G. Andres, *2D Mater.*, 2015, **2**, 011002.

153 A. Favron, E. Gaufres, F. Fossard, A.-L. Phaneuf-Lheureux, N. Y. W. Tang, P. L. Levesque, A. Loiseau, R. Leonelli, S. Francoeur and R. Martel, *Nat. Mater.*, 2015, **14**, 826–832.
154 W. Gaoxue, J. S. William, P. Ravindra and P. K. Shashi, *2D Mater.*, 2016, **3**, 025011.
155 M. T. Edmonds, A. Tadich, A. Carvalho, A. Ziletti, K. M. O'Donnell, S. P. Koenig, D. F. Coker, B. Özyilmaz, A. H. C. Neto and M. S. Fuhrer, *ACS Appl. Mater. Interfaces*, 2015, **7**, 14557–14562.
156 Z. Guo, H. Zhang, S. Lu, Z. Wang, S. Tang, J. Shao, Z. Sun, H. Xie, H. Wang, X.-F. Yu and P. K. Chu, *Adv. Funct. Mater.*, 2015, **25**, 6996–7002.
157 J. Kang, S. A. Wells, J. D. Wood, J.-H. Lee, X. Liu, C. R. Ryder, J. Zhu, J. R. Guest, C. A. Husko and M. C. Hersam, *Proc. Natl. Acad. Sci.*, 2016, 11688–11693.
158 R. J. Smith, P. J. King, M. Lotya, C. Wirtz, U. Khan, S. De, A. O'Neill, G. S. Duesberg, J. C. Grunlan, G. Moriarty, J. Chen, J. Wang, A. I. Minett, V. Nicolosi and J. N. Coleman, *Adv. Mater.*, 2011, **23**, 3944–3948.
159 E. Varrla, C. Backes, K. R. Paton, A. Harvey, Z. Gholamvand, J. McCauley and J. N. Coleman, *Chem. Mater.*, 2015, **27**, 1129–1139.
160 J. R. Brent, A. K. Ganguli, V. Kumar, D. Lewis, P. D. McNaughter, P. Obrien, P. Sabherwal and A. A. Tedstone, *RSC Adv.*, 2016, 86955–86958.
161 Z. Zhu, J. Guan and D. Tomanek, *Nano Lett.*, 2015, **15**, 6042–6046.
162 C. Kamal and M. Ezawa, *Phys. Rev. B: Condens. Matter Mater. Phys.*, 2015, **91**, 085423.
163 G. Wang, R. Pandey and S. P. Karna, *ACS Appl. Mater. Interfaces*, 2015, **7**, 11490–11496.
164 Y. Wang and Y. Ding, *Nanoscale Res. Lett.*, 2015, **10**, 1–10.
165 Y.-p. Wang, W.-x. Ji, C.-w. Zhang, P. Li, F. Li, M.-j. Ren, X.-L. Chen, M. Yuan and P.-j. Wang, *Sci. Rep.*, 2016, **6**, 20342.
166 M. Zhao, X. Zhang and L. Li, *Sci. Rep.*, 2015, **5**, 16108.
167 S. Zhang, Z. Yan, Y. Li, Z. Chen and H. Zeng, *Angew. Chem., Int. Ed.*, 2015, **54**, 3112–3115.
168 O. U. Akturk, V. O. Ozcelik and S. Ciraci, *Phys. Rev. B: Condens. Matter Mater. Phys.*, 2015, **91**, 235446.
169 H.-S. Tsai, C.-W. Chen, C.-H. Hsiao, H. Ouyang and J.-H. Liang, *Chem. Commun.*, 2016, **52**, 8409–8412.
170 H. Cao, Z. Yu and P. Lu, *Superlattices Microstruct.*, 2015, **86**, 501–507.
171 Z. Li, W. Xu, Y. Yu, H. Du, K. Zhen, J. Wang, L. Luo, H. Qiu and X. Yang, *J. Mater. Chem. C*, 2016, **4**, 362–370.
172 K. Luo, S. Chen and C. Duan, *Sci. China: Phys., Mech. Astron.*, 2015, **58**, 87301.
173 H.-S. Tsai, C.-C. Lai, C.-H. Hsiao, H. Medina, T.-Y. Su, H. Ouyang, T.-H. Chen, J.-H. Liang and Y.-L. Chueh, *ACS Appl. Mater. Interfaces*, 2015, **7**, 13723–13727.
174 H.-S. Tsai, S.-W. Wang, C.-H. Hsiao, C.-W. Chen, H. Ouyang, Y.-L. Chueh, H.-C. Kuo and J.-H. Liang, *Chem. Mater.*, 2016, **28**, 425–429.
175 T. Lei, C. Liu, J.-L. Zhao, J.-M. Li, Y.-P. Li, J.-O. Wang, R. Wu, H.-J. Qian, H.-Q. Wang and K. Ibrahim, *J. Appl. Phys.*, 2016, **119**, 015302.
176 C. Huo, Z. Yan, X. Song and H. Zeng, *Sci. Bull.*, 2015, **60**, 1994–2008.
177 Y.-p. Wang, C.-w. Zhang, W.-x. Ji, R.-w. Zhang, P. Li, P.-j. Wang, M.-j. Ren, X.-l. Chen and M. Yuan, *J. Phys. D: Appl. Phys.*, 2016, **49**, 055305.
178 A. N. Alexandrova, A. I. Boldyrev, H.-J. Zhai and L.-S. Wang, *Coord. Chem. Rev.*, 2006, **250**, 2811–2866.
179 A. Quandt and I. Boustani, *ChemPhysChem*, 2005, **6**, 2001–2008.

180 B. Peng, H. Zhang, H. Shao, Y. Xu, R. Zhang and H. Zhua, *J. Mater. Chem. C*, 2016, **4**, 3592–3598.
181 H. Sun, Q. Li and X. G. Wan, *Phys. Chem. Chem. Phys.*, 2016, **18**, 14927–14932.
182 T. Tsafack and B. I. Yakobson, *Phys. Rev. B*, 2016, **93**, 165434.
183 Y. Zhao, S. Zeng and J. Ni, *Phys. Rev. B*, 2016, **93**, 014502.
184 E. S. Penev, A. Kutana and B. I. Yakobson, *Nano Lett.*, 2016, **16**, 2522–2526.
185 X.-F. Zhou, X. Dong, A. R. Oganov, Q. Zhu, Y. Tian and H.-T. Wang, *Phys. Rev. Lett.*, 2014, **112**, 085502.
186 D. Yu, K. Goh, H. Wang, L. Wei, W. Jiang, Q. Zhang, L. Dai and Y. Chen, *Nat. Nanotechnol.*, 2014, **9**, 555–562.
187 W.-L. Li, Y.-F. Zhao, H.-S. Hu, J. Li and L.-S. Wang, *Angew. Chem., Int. Ed.*, 2014, **53**, 5540–5545.
188 Z. A. Piazza, H.-S. Hu, W.-L. Li, Y.-F. Zhao, J. Li and L.-S. Wang, *Nat. Commun.*, 2014, **5**, 3113.
189 X.-B. Li, S.-Y. Xie, H. Zheng, W. Q. Tian and H.-B. Sun, *Nanoscale*, 2015, **7**, 18863–18871.
190 L.-S. Wang, *Int. Rev. Phys. Chem.*, 2016, **35**, 69–142.
191 C.-S. Liu, X. Wang, X.-J. Ye, X. Yan and Z. Zeng, *J. Chem. Phys.*, 2014, **141**, 194306.
192 H. R. Jiang, Z. Lu, M. C. Wu, F. Ciucci and T. S. Zhao, *Nano Energy*, 2016, **23**, 97–104.
193 B. Feng, J. Zhang, Q. Zhong, W. Li, S. Li, H. Li, P. Cheng, S. Meng, L. Chen and K. Wu, *Nat. Chem.*, 2016, **8**, 563–568.
194 A. J. Mannix, X.-F. Zhou, B. Kiraly, J. D. Wood, D. Alducin, B. D. Myers, X. Liu, B. L. Fisher, U. Santiago, J. R. Guest, M. J. Yacaman, A. Ponce, A. R. Oganov, M. C. Hersam and N. P. Guisinger, *Science*, 2015, **350**, 1513–1516.
195 Z. Zhang, Y. Yang, G. Gao and B. I. Yakobson, *Angew. Chem., Int. Ed.*, 2015, **54**, 13022–13026.
196 R. D. Dewhurst, R. Claessen and H. Braunschweig, *Angew. Chem., Int. Ed.*, 2016, **55**, 4866–4868.
197 Y. Jiao, F. Ma, J. Bell, A. Bilic and A. Du, *Angew. Chem., Int. Ed.*, 2016, 10448–10451.
198 J. Yuan, L. W. Zhang and K. M. Liew, *RSC Adv.*, 2015, **5**, 74399–74407.
199 K. L. Marsh, M. Souliman and R. B. Kaner, *Chem. Commun.*, 2015, **51**, 187–190.
200 Y. Stehle, H. M. Meyer, R. R. Unocic, M. Kidder, G. Polizos, P. G. Datskos, R. Jackson, S. N. Smirnov and I. V. Vlassiouk, *Chem. Mater.*, 2015, **27**, 8041–8047.
201 M. H. Khan, Z. Huang, F. Xiao, G. Casillas, Z. Chen, P. J. Molino and H. K. Liu, *Sci. Rep.*, 2015, **5**, 7743.
202 S. M. Kim, A. Hsu, M. H. Park, S. H. Chae, S. J. Yun, J. S. Lee, D.-H. Cho, W. Fang, C. Lee, T. Palacios, M. Dresselhaus, K. K. Kim, Y. H. Lee and J. Kong, *Nat. Commun.*, 2015, **6**, 8662.
203 S. Dai, Q. Ma, M. K. Liu, T. Andersen, Z. Fei, M. D. Goldflam, M. Wagner, K. Watanabe, T. Taniguchi, M. Thiemens, F. Keilmann, G. C. A. M. Janssen, S. E. Zhu, P. Jarillo-Herrero, M. M. Fogler and D. N. Basov, *Nat. Nanotechnol.*, 2015, **10**, 682–686.
204 A. Poddubny, I. Iorsh, P. Belov and Y. Kivshar, *Nat. Photonics*, 2013, **7**, 948–957.
205 S. Dai, Z. Fei, Q. Ma, A. S. Rodin, M. Wagner, A. S. McLeod, M. K. Liu, W. Gannett, W. Regan, K. Watanabe, T. Taniguchi, M. Thiemens,

G. Dominguez, A. H. C. Neto, A. Zettl, F. Keilmann, P. Jarillo-Herrero, M. M. Fogler and D. N. Basov, *Science*, 2014, **343**, 1125–1129.
206 O. Cretu, Y.-C. Lin and K. Suenaga, *Nano Lett.*, 2014, **14**, 1064–1068.
207 H. Cun, M. Iannuzzi, A. Hemmi, J. Osterwalder and T. Greber, *Acs Nano*, 2014, **8**, 7423–7431.
208 N. Alem, R. Erni, C. Kisielowski, M. D. Rossell, W. Gannett and A. Zettl, *Phys. Rev. B: Condens. Matter Mater. Phys.*, 2009, **80**, 155425.
209 L. Makinistian and E. A. Albanesi, *Phys. Rev. B: Condens. Matter Mater. Phys.*, 2006, **74**, 045206.
210 F. Jamali-Sheini, M. Cheraghizade and R. Yousefi, *Sol. Energy Mater. Sol. Cells*, 2016, **154**, 49–56.
211 P. Sinsermsuksakul, L. Sun, S. W. Lee, H. H. Park, S. B. Kim, C. Yang and R. G. Gordon, *Adv. Energy Mater.*, 2014, **4**, 1400496.
212 J. R. Brent, D. J. Lewis, T. Lorenz, E. A. Lewis, N. Savjani, S. J. Haigh, G. Seifert, B. Derby and P. O'Brien, *J. Am. Chem. Soc.*, 2015, **137**, 12689–12696.
213 S. G. Hickey, C. Waurisch, B. Rellinghaus and A. Eychmüller, *J. Am. Chem. Soc.*, 2008, **130**, 14978–14980.
214 J. Y. Kim and S. M. George, *J. Phys. Chem. C*, 2010, **114**, 17597–17603.
215 N. Koteeswara Reddy, M. Devika and E. S. R. Gopal, *Crit. Rev. Solid State Mater. Sci.*, 2015, **40**, 359–398.
216 T. Jiang and G. A. Ozin, *J. Mater. Chem.*, 1998, **8**, 1099–1108.
217 H. Zhang, C. Xia, X. Zhao, T. Wang and J. Li, *Appl. Surf. Sci.*, 2015, **356**, 1200–1206.
218 T. Lorenz, J.-O. Joswig and G. Seifert, *Semicond. Sci. Technol.*, 2014, **29**, 064006.
219 H. Wei, Y. Su, S. Chen, Y. Lin, Z. Yang, X. Chen and Y. Zhang, *J. Mater. Chem.*, 2011, **21**, 12605–12608.
220 J. Yu, C.-Y. Xu, Y. Li, F. Zhou, X.-S. Chen, P.-A. Hu and L. Zhen, *Sci. Rep.*, 2015, **5**, 17109.
221 K. S. Novoselov, A. K. Geim, S. V. Morozov, D. Jiang, Y. Zhang, S. V. Dubonos, I. V. Grigorieva and A. A. Firsov, *Science*, 2004, **306**, 666–669.
222 D. J. Lewis, P. Kevin, O. Bakr, C. A. Muryn, M. A. Malik and P. O'Brien, *Inorg. Chem. Front.*, 2014, **1**, 577–598.
223 S. Najmaei, Z. Liu, W. Zhou, X. Zou, G. Shi, S. Lei, B. I. Yakobson, J.-C. Idrobo, P. M. Ajayan and J. Lou, *Nat. Mater.*, 2013, **12**, 754–759.
224 G. Su, D. De, V. G. Hadjiev and H. Peng, in *Micro- and Nanotechnology Sensors, Systems, and Applications Vi*, ed. T. George, M. S. Islam and A. K. Dutta, 2014.
225 S. Sohila, M. Rajalakshmi, C. Muthamizhchelvan, S. Kalavathi, C. Ghosh, R. Divakar, C. N. Venkiteswaran, N. G. Muralidharan, A. K. Arora and E. Mohandas, *Mater. Lett.*, 2011, **65**, 1148–1150.
226 J. Ning, K. Men, G. Xiao, L. Wang, Q. Dai, B. Zou, B. Liu and G. Zou, *Nanoscale*, 2010, **2**, 1699–1703.
227 M. D. Khan, J. Akhtar, M. A. Malik, M. Akhtar and N. Revaprasadu, *New J. Chem.*, 2015, **39**, 9569–9574.
228 X. Zhang, L. Yang, Y. Jiang, B.-B. Yu, Y.-G. Zou, Y. Fang, J.-S. Hu and L.-J. Wan, *Chem. – Asian J.*, 2013, **8**, 2483–2488.
229 H. Zhu, D. Yang, Y. Ji, H. Zhang and X. Shen, *J. Mater. Sci.*, 2005, **40**, 591–595.
230 Y. Zhang, J. Lu, S. Shen, H. Xu and Q. Wang, *Chem. Commun.*, 2011, **47**, 5226–5228.

231 S. M. Herron, J. T. Tanskanen, K. E. Roelofs and S. F. Bent, *Chem. Mater.*, 2014, **26**, 7106–7113.
232 W. Gao, C. Wu, M. Cao, J. Huang, L. Wang and Y. Shen, *J. Alloys Compd.*, 2016, **688**(Part A), 668–674.
233 A. de Kergommeaux, M. Lopez-Haro, S. Pouget, J.-M. Zuo, C. Lebrun, F. Chandezon, D. Aldakov and P. Reiss, *J. Am. Chem. Soc.*, 2015, **137**, 9943–9952.
234 W. Sun, X. Rui, D. Yang, Z. Sun, B. Li, W. Zhang, Y. Zong, S. Madhavi, S. Dou and Q. Yan, *ACS Nano*, 2015, **9**, 11371–11381.
235 J. Wang, G. Lian, Z. Xu, C. Fu, Z. Lin, L. Li, Q. Wang, D. Cui and C.-P. Wong, *ACS Appl. Mater Interfaces*, 2016, **8**, 9545–9551.
236 J. Yu, C.-Y. Xu, F.-X. Ma, S.-P. Hu, Y.-W. Zhang and L. Zhen, *ACS Appl. Mater. Interfaces*, 2014, **6**, 22370–22377.
237 R. Wei, J. Hu, T. Zhou, X. Zhou, J. Liu and J. Li, *Acta Mater.*, 2014, **66**, 163–171.
238 D. Gao, Q. Xue, X. Mao, M. Xue, S. Shi and D. Xue, *CrystEngComm*, 2014, **16**, 7876–7880.
239 Y. Liu, H. Kang, L. Jiao, C. Chen, K. Cao, Y. Wang and H. Yuan, *Nanoscale*, 2015, **7**, 1325–1332.
240 Z. X. Huang, Y. Wang, J. I. Wong and H. Y. Yang, *2D Mater.*, 2015, **2**, 024010.
241 B. Yang, X. Zuo, P. Chen, L. Zhou, X. Yang, H. Zhang, G. Li, M. Wu, Y. Ma, S. Jin and X. Chen, *ACS Appl. Mater. Interfaces*, 2015, **7**, 137–143.
242 Y.-M. Han, J. Zhao, M. Zhou, X.-X. Jiang, H.-Q. Leng and L.-F. Li, *J. Mater. Chem. A*, 2015, **3**, 4555–4559.
243 A. A. Tedstone, D. J. Lewis and P. O'Brien, *Chem. Mater.*, 2016, **28**, 1965–1974.
244 G. Bai, S. Yuan, Y. Zhao, Z. Yang, S. Y. Choi, Y. Chai, S. F. Yu, S. P. Lau and J. Hao, *Adv. Mater.*, 2016, 7472–7477.
245 S.-H. Su, Y.-T. Hsu, Y.-H. Chang, M.-H. Chiu, C.-L. Hsu, W.-T. Hsu, W.-H. Chang, J.-H. He and L.-J. Li, *Small*, 2014, **10**, 2589–2594.
246 G. Wang, C. Robert, A. Suslu, B. Chen, S. Yang, S. Alamdari, I. C. Gerber, T. Amand, X. Marie, S. Tongay and B. Urbaszek, *Nat. Commun.*, 2015, **6**, 10110.
247 D. Jariwala, V. K. Sangwan, L. J. Lauhon, T. J. Marks and M. C. Hersam, *ACS Nano*, 2014, **8**, 1102–1120.
248 J. H. Han, S. Lee and J. Cheon, *Chem. Soc. Rev.*, 2013, **42**, 2581–2591.
249 L. Sun, Z. Lin, J. Peng, J. Weng, Y. Huang and Z. Luo, *Sci. Rep.*, 2014, **4**, 4794.
250 T. Ludwig, L. Guo, P. McCrary, Z. Zhang, H. Gordon, H. Quan, M. Stanton, R. M. Frazier, R. D. Rogers, H.-T. Wang and C. H. Turner, *Langmuir*, 2015, **31**, 3644–3652.
251 L. Yuwen, H. Yu, X. Yang, J. Zhou, Q. Zhang, Y. Zhang, Z. Luo, S. Su and L. Wang, *Chem. Commun.*, 2016, **52**, 529–532.
252 A. O'Neill, U. Khan and J. N. Coleman, *Chem. Mater.*, 2012, **24**, 2414–2421.
253 V. Nicolosi, M. Chhowalla, M. G. Kanatzidis, M. S. Strano and J. N. Coleman, *Science*, 2013, **340**, 6139.
254 G. Cunningham, M. Lotya, C. S. Cucinotta, S. Sanvito, S. D. Bergin, R. Menzel, M. S. P. Shaffer and J. N. Coleman, *ACS Nano*, 2012, **6**, 3468–3480.
255 K. R. Paton, E. Varrla, C. Backes, R. J. Smith, U. Khan, A. O'Neill, C. Boland, M. Lotya, O. M. Istrate, P. King, T. Higgins, S. Barwich, P. May, P. Puczkarski, I. Ahmed, M. Moebius, H. Pettersson, E. Long, J. Coelho, S. E. O'Brien, E. K. McGuire, B. M. Sanchez, G. S. Duesberg, N. McEvoy,

T. J. Pennycook, C. Downing, A. Crossley, V. Nicolosi and J. N. Coleman, *Nat. Mater.*, 2014, **13**, 624–630.
256 D. Chakravarty and D. J. Late, *Eur. J. Inorg. Chem.*, 2015, 1973–1980.
257 D. Li, M. B. Muller, S. Gilje, R. B. Kaner and G. G. Wallace, *Nat. Nanotechnol.*, 2008, **3**, 101–105.
258 F. Cui, C. Wang, X. Li, G. Wang, K. Liu, Z. Yang, Q. Feng, X. Liang, Z. Zhang, S. Liu, Z. Lei, Z. Liu, H. Xu and J. Zhang, *Adv. Mater.*, 2016, **28**, 5019–5024.
259 M.-Y. Li, Y. Shi, C.-C. Cheng, L.-S. Lu, Y.-C. Lin, H.-L. Tang, M.-L. Tsai, C.-W. Chu, K.-H. Wei, J.-H. He, W.-H. Chang, K. Suenaga and L.-J. Li, *Science*, 2015, **349**, 524–528.
260 W. Jung, S. Lee, D. Yoo, S. Jeong, P. Miró, A. Kuc, T. Heine and J. Cheon, *J. Am. Chem. Soc.*, 2015, **137**, 7266–7269.
261 P. D. Antunez, D. H. Webber and R. L. Brutchey, *Chem. Mater.*, 2013, **25**, 2385–2387.
262 B. Mahler, V. Hoepfner, K. Liao and G. A. Ozin, *J. Am. Chem. Soc.*, 2014, **136**, 14121–14127.
263 N. Savjani, E. A. Lewis, M. A. Bissett, J. R. Brent, R. A. W. Dryfe, S. J. Haigh and P. O'Brien, *Chem. Mater.*, 2016, **28**, 657–664.
264 E. Devers, P. Afanasiev, B. Jouguet and M. Vrinat, *Catal. Lett.*, 2002, **82**, 13–17.
265 Y. Peng, Z. Meng, C. Zhong, J. Lu, W. Yu, Y. Jia and Y. Qian, *Chem. Lett.*, 2001, **30**, 772–773.
266 S. Wang, G. Li, G. Du, X. Jiang, C. Feng, Z. Guo and S.-J. Kim, *Chin. J. Chem. Eng.*, 2010, **18**, 910–913.
267 A. Pospischil and T. Mueller, *Appl. Sci.*, 2016, **6**, 78.
268 H. Tian, M. L. Chin, S. Najmaei, Q. Guo, F. Xia, H. Wang and M. Dubey, *Nano Res.*, 2016, **9**, 1543–1560.
269 B. Peng, P. K. Ang and K. P. Loh, *Nano Today*, 2015, **10**, 128–137.
270 R. Lv, J. A. Robinson, R. E. Schaak, D. Sun, Y. Sun, T. E. Mallouk and M. Terrones, *Acc. Chem. Res.*, 2015, **48**, 56–64.
271 M. Bernardi, M. Palummo and J. C. Grossman, *Nano Lett.*, 2013, **13**, 3664–3670.
272 N. Harada, S. Sato and N. Yokoyama, *J. Appl. Phys.*, 2014, **115**, 034505.
273 H.-M. Li, D.-Y. Lee, M. S. Choi, D. Qu, X. Liu, C.-H. Ra and W. J. Yoo, *Sci. Rep.*, 2014, **4**, 4041.
274 T. Paul, S. Ghatak and A. Ghosh, *Nanotechnology*, 2016, **27**, 125706.
275 M.-Y. Tsai, A. Tarasov, Z. R. Hesabi, H. Taghinejad, P. M. Campbell, C. A. Joiner, A. Adibi and E. M. Vogel, *ACS Appl. Mater. Interfaces*, 2015, **7**, 12850–12855.
276 J. Wang, D. Rhodes, S. Feng, M. A. T. Nguyen, K. Watanabe, T. Taniguchi, T. E. Mallouk, M. Terrones, L. Balicas and J. Zhu, *Appl. Phys. Lett.*, 2015 **106**, 152104.
277 Y. Kobayashi, S. Mori, Y. Maniwa and Y. Miyata, *Nano Res.*, 2015, **8**, 3261–3271.
278 W. Zhang, C.-P. Chuu, J.-K. Huang, C.-H. Chen, M.-L. Tsai, Y.-H. Chang, C.-T. Liang, Y.-Z. Chen, Y.-L. Chueh, J.-H. He, M.-Y. Chou and L.-J. Li, *Sci. Rep.*, 2014, **4**, 3826.
279 S. J. Kang, G.-H. Lee, Y.-J. Yu, Y. Zhao, B. Kim, K. Watanabe, T. Taniguchi, J. Hone, P. Kim and C. Nuckolls, *Adv. Funct. Mater.*, 2014, **24**, 5157–5163.
280 Y. Wang and Y. Ding, *Phys. Chem. Chem. Phys.*, 2015, **17**, 27769–27776.
281 K. E. Aretouli, D. Tsoutsou, P. Tsipas, J. Marquez-Velasco, S. Aminalragia Giamini, N. Kelaidis, V. Psycharis and A. Dimoulas, *ACS Appl. Mater. Interfaces*, 2016, 23222–23229.

282 D. Voiry, J. Yang and M. Chhowalla, *Adv. Mater.*, 2016, **28**, 6197–6206.
283 F.-X. Xiao, J. Miao and B. Liu, *J. Am. Chem. Soc.*, 2014, **136**, 1559–1569.
284 Y. Hu and D. H. C. Chua, *Sci. Rep.*, 2016, **6**, 28088.
285 D. Deng, K. S. Novoselov, Q. Fu, N. Zheng, Z. Tian and X. Bao, *Nat. Nanotechnol.*, 2016, **11**, 218–230.
286 J. M. Woods, Y. Jung, Y. Xie, W. Liu, Y. Liu, H. Wang and J. J. Cha, *ACS Nano*, 2016, **10**, 2004–2009.
287 X. Sun, H. Deng, W. Zhu, Z. Yu, C. Wu and Y. Xie, *Angew. Chem., Int. Ed.*, 2016, **55**, 1704–1709.
288 L. Liao, J. Zhu, X. Bian, L. Zhu, M. D. Scanlon, H. H. Girault and B. Liu, *Adv. Funct. Mater.*, 2013, **23**, 5326–5333.
289 X. Zheng, J. Xu, K. Yan, H. Wang, Z. Wang and S. Yang, *Chem. Mater.*, 2014, **26**, 2344–2353.
290 Y.-H. Chang, C.-T. Lin, T.-Y. Chen, C.-L. Hsu, Y.-H. Lee, W. Zhang, K.-H. Wei and L.-J. Li, *Adv. Mater.*, 2013, **25**, 756–760.
291 J. Wang, J. Liu, D. Chao, J. Yan, J. Lin and Z. X. Shen, *Adv. Mater.*, 2014, **26**, 7162–7169.
292 J. Shi, X. Zhou, G.-F. Han, M. Liu, D. Ma, J. Sun, C. Li, Q. Ji, Y. Zhang, X. Song, X.-Y. Lang, Q. Jiang, Z. Liu and Y. Zhang, *Adv. Mater. Interfaces*, 2016, 1600332.
293 L. David, R. Bhandavat and G. Singh, *ACS Nano*, 2014, **8**, 1759–1770.
294 L. Yu, Y.-H. Lee, X. Ling, E. J. G. Santos, Y. C. Shin, Y. Lin, M. Dubey, E. Kaxiras, J. Kong, H. Wang and T. Palacios, *Nano Lett.*, 2014, **14**, 3055–3063.
295 L. Samad, S. M. Bladow, Q. Ding, J. Zhuo, R. M. Jacobberger, M. S. Arnold and S. Jin, *ACS Nano*, 2016, **10**, 7039–7046.
296 J. Wang, Q. Yao, C.-W. Huang, X. Zou, L. Liao, S. Chen, Z. Fan, K. Zhang, W. Wu, X. Xiao, C. Jiang and W.-W. Wu, *Adv. Mater.*, 2016, **37**, 8302–8308.
297 C. Xia, B. Xue, T. Wang, Y. Peng and Y. Jia, *Appl. Phys. Lett.*, 2015, **107**, 193107.
298 G. V. Naik, V. M. Shalaev and A. Boltasseva, *Adv. Mater.*, 2013, **25**, 3264–3294.
299 Y. Zhang, Y.-W. Tan, H. L. Stormer and P. Kim, *Nature*, 2005, **438**, 201–204.

Colloidal quantum dots solar cells

Richard A. Taylor*[a] and Karthik Ramasamy[b]
DOI: 10.1039/9781782620358-00142

Solar cells fabricated from conventional bulk semiconductors such as silicon or gallium arsenide are approaching the physical limit of solar power conversion efficiency. It is believed that adapting the solar cell to the solar spectrum is one suitable approach to improving conversion efficiencies. On that basis nanostructured solar cells are attracting significant attention as promising alternatives because on that length scale the associated optoelectronic properties are improved. This chapter will explore the application of colloidal quantum dots in photovoltaic devices. Premised on historical developments their electronic structures arising from quantum confinement that results from reduced dimensionality are discussed. Optoelectronic properties are discussed based on effective methods of synthesis and how they enable control of chemical and physical properties. There is a focus on developments from fabrication methods to device physics looking at several types and performance characteristics in terms of charge carrier properties and their effect on efficiencies.

1 Introduction, historical background of colloidal quantum dots

The field of photovoltaics has been rapidly expanding since its emergence in the 1950s starting with the fabrication of monocrystalline silicon wafer single junction cells.[1] This is partly to overcome one of the foremost challenges of reducing the cost per watt of delivered solar electricity. However, in order to overcome the theoretical efficiencies for single crystalline silicon solar cells of 26–28%, the field has developed to find effective alternatives and the result is several categories of photovoltaic (PV) devices with third generation quantum dot (QD) based cells gaining much attention. One basis of low efficiencies in conventional microstructured (bulk) single junction solar cells is that sub *bandgap*, E_g photons are not absorbed while supra bandgap photons produce hot-charge carriers and thermalize the excess energy as heat. Therefore, novel materials or structures with tunable bandgap to match the spectral distribution of solar spectrum are crucial. Quantum dot based solar cells offer the advantage of tunable bandgap, which can adapt the solar cell to the solar spectrum to utilize more energy producing extra charge carriers, increasing conductivity and photoconversion. Additionally, they can be processed to create junctions (multiple) on inexpensive substrates such as plastics, glass or metal sheets and can easily be combined with organic polymers and dyes. PV devices of this special class of semiconductors emerged in the late 1990s with efficiencies below 1%. However, over the years with improved device designs and greater understanding of material properties, reported efficiencies now range from range from 4.0–8.6%.[2]

[a]*The University of the West Indies, Department of Chemistry, St. Augustine, Trinidad and Tobago. E-mail: richard.taylor@sta.uwi.edu*
[b]*UbiQD, LLC. Los Alamos NM-87505, USA*

Colloidal quantum dots (*CQDs*), inorganic semiconductor nanocrystals (<10 nm) typically capped with surfactant organic molecules and dispersed in solution have provided a powerful platform for the development of several classes of cost-effective, solution-processed optoelectronic devices over the past decade, including photovoltaic cells, photodetectors and light-emission devices.[3] A unique advantage of CQDs is the *quantum confinement effect*[4] where, optoelectronic properties are dependent on their size and shape. In effect, the bandgap increases with a decrease in size of the quantum dot, and this tunability allows for fabrication of nanostructured solar cells that more effectively absorb energy from the solar spectrum. Additionally, quantum dots have large intrinsic dipole moments ideal for rapid charge separation in solar cells. They have been found to exhibit multiple exciton generation (MEG), by emitting up to three electrons per photon and theoretically could increase photovoltaic efficiency from 20 to 65%.[5]

The state-of-play in photovoltaics is strongly focused on several types of fabricated quantum dots based solar cell devices including doped and undoped Schottky junction cells, heterojunction cells (p–n and p–i–n junctions), multi-junction cells, quantum dot-sensitized cells and hybrid cells. In this regard, the overall objective of this chapter is to present an overview of the current state-of-the-art of this expanding field premised on the fundamental chemistry and physics of CQDs including, chemical and surface properties, size-dependent optoelectronic properties, methods of synthesis, description of cell architectures, device fabrication and performance.

2 Semiconductors: dimensionality and quantum confined structures

2.1 Electronic band structure

The electro-optical properties of semiconductor solids are dependent on their band structure, which is based on a quantum mechanical treatment of charger carriers (electrons and holes) described using *Band Theory*.[6] The theory is specifically premised on the energy of a nearly-free electron travelling through the crystal with a periodic potential or a *wave vector, k* based on the periodicity of the lattice (nuclei) described using the *Bloch Theorem*. The result is the formation of two distinct energy bands as seen in Fig. 1, the lower called the *valence band*, filled with electrons and the empty upper called the *conduction band*.

When the material is excited either by increase in temperature or irradiation, electrons that have absorbed energy equivalent to the bandgap are able to transition into the conduction band leaving behind holes in the valence band. The mobility of electrons and holes (excitons: electron–hole pairs) is a measure of the degree of conductivity in materials. These interband transitions, absorption (a) and luminescence (b), whether direct or indirect are specific to various semiconductor materials and uniquely define their optoelectronic properties.

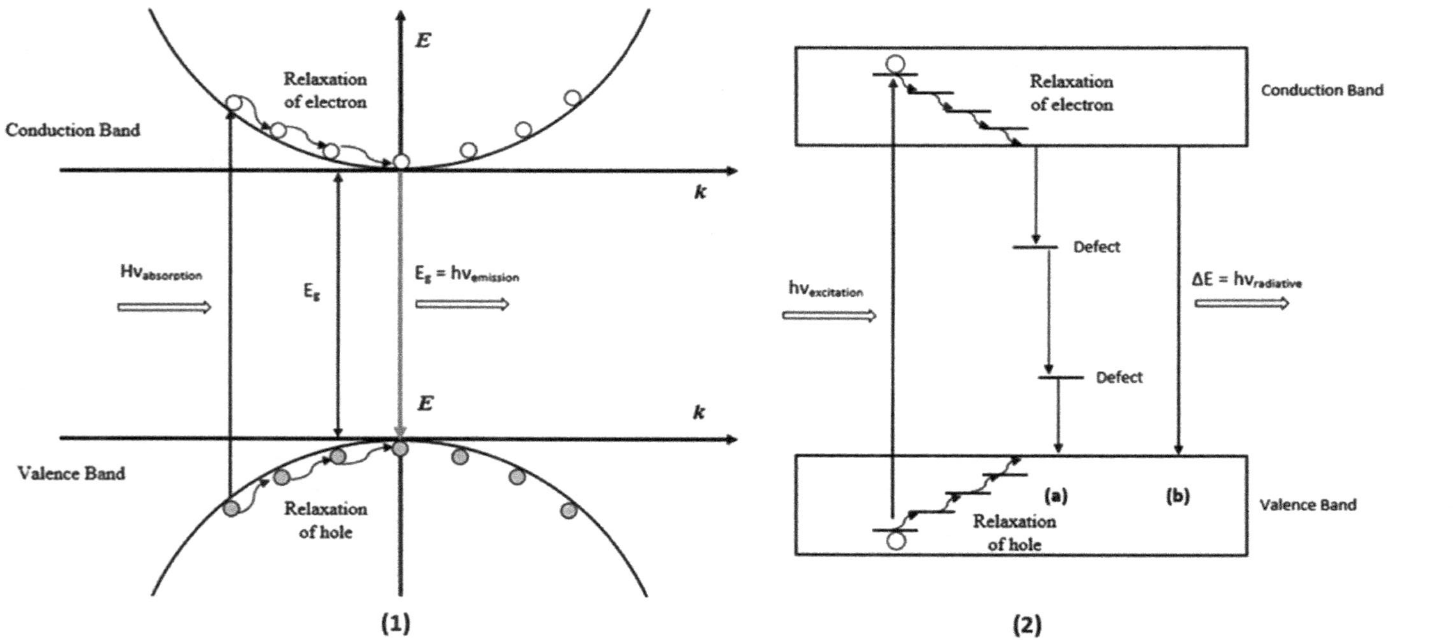

Fig. 1 Energy band diagrams (1) *E vs. k* graph and (2) flat band showing exciton relaxation followed by recombination for (a) non-radiative and (b) radiative emission due to defect states.

2.2 Quantum confinement

The electro-optical properties of semiconductor solids display size-dependence as a consequence of quantum confinement.[4] This phenomenon, typical for quantum dots, is premised on the Heisenberg uncertainty principle in which electrons and holes are confined to a specific region of space, comparable to the size of the dot, the *Bohr exciton radius*. As a consequence, quantum confined structures are generally classified in terms of their dimensionality (size and shape). In bulk solids, the electrons a free to move in all three directions, *i.e.* they have three degrees of freedom and display three-dimensional (3-D) physics. However, quantum dots have 3-D confinement and are described as 0-D nanostructures. Similarly, *quantum wires* and *quantum wells* as illustrated in Fig. 2 have 2-D and 1-D confinement respectively, and their *density of states* (number of available energy states available to the excited electron) and are dependent on the dimensionality.[6,7]

At these dimensions where the quantum dots typically contain on the order of 10^3–10^6 atoms the spatial confinement of excitons and the electronic structure become intermediate between that of an atom and a bulk semiconductor, *i.e.* the continuous bands of the bulk now become discrete in the nanostructure and the magnitude of the bandgap increases. As such for quantum dots, size-dependence is a unique feature that is observed and fully exploited in the optical properties, where there is red-shift in absorption band edge (bandgap, $E_{g(\mathrm{QD})}$) relative to bulk bandgap, $E_{g(\mathrm{bulk})}$ as particle size increases as illustrated in Fig. 3. Brus summarises this accordingly:[8]

$$E_{g(\mathrm{QD})} = E_{g(\mathrm{bulk})} - \frac{\hbar^2\pi^2}{2R^2}\left\{\frac{1}{m_e^*} + \frac{1}{m_h^*}\right\} - \frac{1.8e^2}{R\varepsilon} \qquad (1)$$

where, R represents the radius of the nanocrystal, e the elementary charge, ε the bulk dielectric coefficient, m_e^* and m_e^*, the effective mass of electron and hole, respectively. Size-dependent optical properties are very attractive for various applications including light absorbing characteristics in quantum dot solar cells since this tunability offers potential advantages over conventional p-n, tandem, hybrid bulk semiconductor solar cells, since the cells can be fabricated to adapt to varying and specific regions of the solar spectrum and possibly increase conductivity.

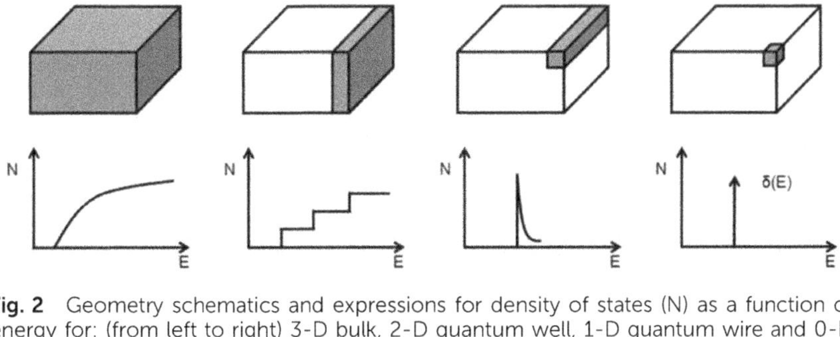

Fig. 2 Geometry schematics and expressions for density of states (N) as a function of energy for: (from left to right) 3-D bulk, 2-D quantum well, 1-D quantum wire and 0-D quantum dot semiconductor structures.[7]

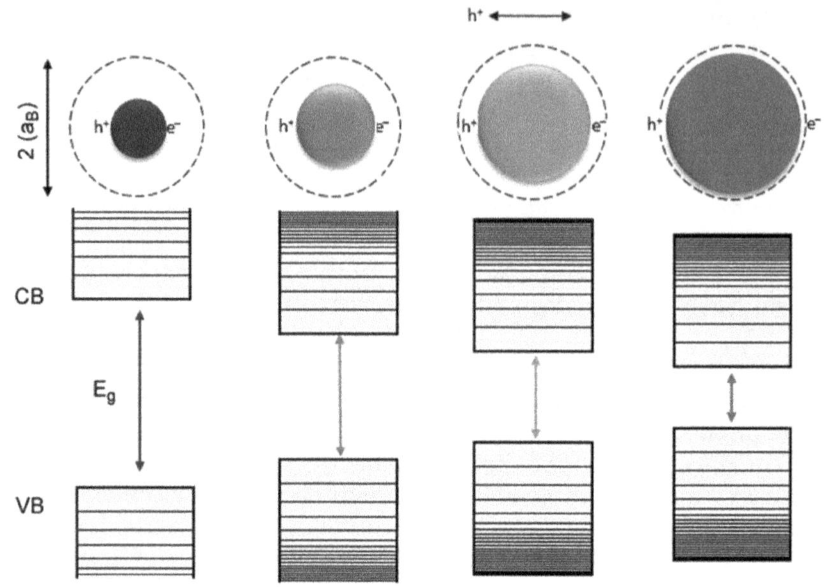

Fig. 3 Change in band gap relative to size and respective absorption or emission colour due to the confinement of electron–hole pair when quantum dot size approaches the Bohr radius.

3 Optoelectronic properties of CQDs

Because the bandgap, interband absorption and luminescence are critical properties responsible for separation of charge carriers in bulk semiconductors and quantum dots an important factor in these radiative processes is the role of intrinsic and extrinsic defect (trapping) states. Overall, the critical parameters that affect conversion efficiencies in CQD solar cells are carrier concentration, mobility and lifetime, trap density, doping density and diffusion length in processed films and these will be discussed later. These are related to device performance parameters including: short-circuit current, J_{sc}, the current through the solar cell when the voltage across the solar cell is zero and is the largest current which may be drawn from the solar cell; open-circuit current, V_{oc}, the maximum voltage available from a solar cell which occurs at zero current; fill-factor, *FF*, the ratio of the maximum power from the solar cell to the product of V_{oc} and J_{oc} which determines the maximum power from a solar cell.[9,10] In order to understand the basis of these properties in CQDs, it is important to probe the interband processes, including the size-dependent absorption and emission, and associated properties such as multiple exciton generation and upconversion/downconversion processes and how they impact photoconductive efficiencies.

3.1 Multiple exciton generation

Multiple exciton generation (MEG) is a form of impact ionization or electron–hole pair multiplication (EHPM) that occurs in semiconductor nanostructures, especially those that exhibit strong quantum confinement. Theoretically, MEG process can boost solar energy conversion efficiency over 60%.[11] Impact ionization is a process by which all or some of the excess energy, at least equal to the bandgap energy of the hot carriers goes to generate additional electron–hole pairs.[12,13] These extra carriers can be extracted from the cell to enhance the photocurrent and thus the overall photovoltaic efficiency. MEG can only occur if the photon energy is above a threshold value, $h\nu_{\text{th}}$ and at least one of the charge carriers created by absorption of a photon has sufficient energy to promote another electron across the bandgap, creating a second exciton. Typically, the average number of excitons created in a QD per photon absorbed, the quantum yield (QY)>100% is a measure of MEG accordingly,[13,14]

$$\text{QY} = \left(\frac{h\nu}{E_g} - 1\right)\eta_{\text{EHPM}} = \frac{(h\nu - E_g)1}{\varepsilon_{\text{EHPM}}} \tag{2}$$

where, η_{EHPM} is EHPM efficiency and $\varepsilon_{\text{EHPM}}$ is the electron–hole pair creation energy, the amount of energy required to produce an additional EHP when the EHPM efficiency is less than 1 (100%). The energy threshold is related to the efficiency by

$$h\nu_{\text{th}} = E_g + \frac{E_g}{\eta_{\text{EHPM}}} \tag{3}$$

The main challenge with determining MEG is based on the timescale of the event. Excitons in QDs have lifetimes on the order 10–100 ns whilst a multi-exciton can undergo rapid recombination on the order 10–100 ps. As such, QY must thus be measured before the MEG-created excitons can undergo this rapid recombination, necessitating the use of ultrafast (subpicosecond) techniques such as transient absorption/emission spectroscopy.[12] In order for MEG to have a large impact on solar energy conversion, the QDs must be incorporated into a suitable energy conversion architecture. The minimum requirements are (1) the QDs must be the absorbing component, (2) the multiexcitons produced within the QDs must be separated prior to *Auger* recombination, and (3) the free charge carriers or excitons must be transported to electron- and hole-accepting contacts.[13] As an example, the efficiency as twice as bulk has been realized in PbSe quantum dots through the MEG process. The CQDs solar cells showing external quantum efficiency (EQE) over 100% was demonstrated using hydrazine treated PbSe quantum dots.[14] Transient absorption spectroscopy was effectively utilized to show MEG for single photon absorption in colloidal PbSe and PbS QDs. Quantum yields of 300% indicate the creation, on average, of three excitons per absorbed photon for PbSe QDs at photon energies four times the QD energy gap.[15] Recently Tomić and co-workers[16] showed that colloidal core/shell CdSe/CdTe type-II QDs lead to (1) strongly bound biexcitonic states that minimize the threshold for the MEG process; (2) increased radiative lifetimes when compared with equivalent core-only QDs (enabling better charge extraction); and (3) while radiative relaxation times are increased by one order of magnitude with introduction of shells, the detrimental Auger electron cooling process is slowed down by three orders of magnitude from ~1 ps to ~1 ns time scale. All those effects acting together can make the colloidal core/shell QD structures promising materials for MEG solar cells.

3.2 Photon upconversion (UC)

Upconversion is a nonlinear process in which two or more long wavelength photons are converted to a shorter wavelength photon.[17] It holds great promise for photovoltaic devices as a means to surpass the Shockley–Queisser efficiency limit, circumventing energy loss by converting two sub-bandgap photons into one supra-bandgap photon, where the PV cell has high light responsivity. This process is possible in heterostructures incorporating two quantum dots of different bandgaps connected *via* a quantum well (rod) acting as a tunnelling barrier.[17,18] This process is based on sequential absorption of two or more photons, involving metastable, long-lived intermediate energy states. The requirements for UC are long-lived excited states, a ladderlike arrangement of energy levels. The first photon generates an electron–hole pair *via* interband absorption in the small band-gap quantum dot, leaving a confined hole and a delocalized electron in the compound semiconductor nanocrystal. The second absorbed photon can lead, either directly or indirectly to further excitation of the hole, enabling it to cross the barrier layer. The direct path is through an intraband absorption of the photon by the confined hole at the lower-energy core (Fig. 4a), while the indirect way is *via* an Auger mediated energy transfer process (Fig. 4b).[19,20]

The Auger process describes a recombination of the second electron–hole pair, generated by absorbing the second photon, while simultaneously allowing non-radiative energy transfer to the confined hole at the small bandgap quantum dot, empowering it to cross the barrier to the higher energy quantum dot. This, in turn, is followed by a radiative recombination with the delocalized electron, producing upconverted luminescence. This system combines the stability of an inorganic crystalline structure, with the spectral tunability afforded by quantum confinement. Since the absorption, emission, and lifetime of the

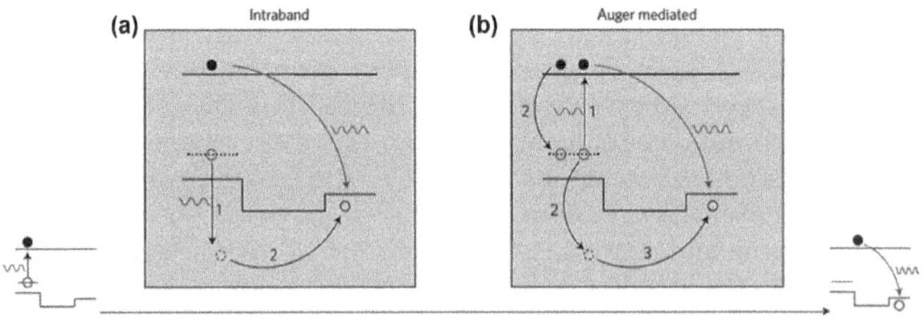

Fig. 4 Mechanism of the upconversion in quantum nanostructures: (a) Direct intraband hole absorption mechanism of upconversion; (b) Auger-mediated upconversion. (Reprinted with permission from Macmillan Publishers Ltd: Nature Nanotechnology (ref. 19); Copyright 2013.

semiconductor nanocrystals can be controlled by variation of their size, shape, as well as composition, upconversion in quantum nanostructures holds prime promise for applications in PV devices.

Recently, Deutsch *et al.*[20] reported a system of upconverting NCs, which incorporates two quantum dots with different bandgaps separated by a tunneling barrier which was able to upconvert 680 nm into 570 nm light. It was shown that these NCs can operate at room temperature, with the dominant mechanism proven to be sequential interband and intraband absorption events. It was also shown by Oron *et al.*[17] that dual near-infrared and visible emitting PbSe/CdSe/CdS heterostructured nanocrystals were able to upconvert a broad range of NIR (1.2 μm) wavelengths to visible emission at room temperature. At saturation, 7 NIR photon pairs are equivalent to the absorption of a single visible photon, a promising attribute of broadband NIR response and band edge tunability lacking in other methods for UC. As such, these NCs can be attractive and promising for use in solar cells that will be able to upconvert the solar NIR spectrum to fit the junction bandgap in solid state devices.

4 CQD chemical properties

Even though optoelectronic properties of colloidal quantum dots are size dependent, they are instrinsically based on their chemical structure and composition with surface properties being critical. As such, quantum dots can be categorised as (1) core type, (2) core/shell type and (3) alloyed. Heterostructured CQDs can be considered as alloyed semiconductors. It has been widely established that the composition of a semiconductor influences the electronic properties. As such, the bandgap of a CQD can be tuned by keeping the size and shape constant and alloying or changing the composition of the core. It has been shown that the relationship between the bandgap and the mole fraction of each component is nonlinear and governed by a bowing parameter dependent on the two binary materials.[3] Tuning the bandgap of alloyed and mulitnary CQDs *via* stoichiometry enables absorption and emission of wavelengths that could not be achieved by the original binary materials. The distribution of the alloy constituents determines whether the alloy is homogeneous or graded.

The most commonly used materials for CQDs are metal chalcogenides (sulphides, selenides and tellurides) such as binary, II–VI and IV–VI, CdS, CdSe, CdTe, ZnS, ZnSe, ZnTe, PbS and PbSe, ternary I–IV–VI, $CuInS_2$/$CuInSe_2$ (CIS), AgInS (AIS), and quaternary, $CuInGaS_2$ (CIGS), $Cu(InGa)Se_2$ (CIGSe), $Cu_2ZnSn(S,Se)_4$ (CZTSSe) and Cu_2FeSnS_4 (CFTS).[21] However, in recent years there have been strong focus in semiconductors consisting of earth abundant and environmentally benign elements due to concerns of toxicity with elements such as lead and cadmium.[22] Recently, Altıntas *et al.*[23] reported a detailed synthetic approach for environmentally friendly and highly efficient InPZnS alloy/ZnS shell quantum dots. Following a systematic study of the ratio and type of the precursors involved, they achieved alloyed core/shell InPZnS/ZnS quantum dots with tunable emission across the visible spectrum, having a record quantum efficiency up to 78% along with a full width at half-maximum as narrow as 45 nm. Using time-resolved photoluminescence spectroscopy they demonstrated the effect of the systematic shell growth in suppressing non-radiative decay channels, with the photoluminescence lifetime prolonged from 20.3 to 50.4 ns. It is envisaged that development of highly efficient and environmentally friendly quantum dots will pave the way for robust, sustainable optoelectronic applications.

4.1 Classification of CQDs

4.1.1 Single core: surface properties, defects and surface modification. Single core CQDs are the simplest and most widely studied. These are typically homostructural of uniform composition and phase comprising an inorganic crystalline core, surface atoms and surface capping/passivating organic ligands (added during synthesis). Modifying the surface chemistry control the optical properties as well as the charge and the energy transfer in assemblies and allow for their integration into thin film solid-state electro-optical devices such as solar cells. The surface atoms do not have complete valence states arising from incomplete bonding (unsaturared/dangling bonds) and due to the high surface-to-volume ratio manifest as surface defects allowing for donor or acceptor electronic trapping states within the bandgap or close to the valence and conduction band edges. These trapping states faciliate non-radiative electron–hole recombination, as shown in Fig. 1 and in particular increase luminescence decay times, decrease quantum yield and efficiency. The surface active capping ligands are often effective in forming quasi-bonds (or covalent bonds) with the surface atoms reducing the density of states associated with these defects. The result is an improvement in emission properties from higher electronic transition rates associated with radiative recombination of electrons and holes.

Even though the most effective passivating ligands that produce good quality, stable colloidal quantum dots with relatively minimal surface defects are long bulky ones such as dodecanthiol, oleylamine, oleic acid and trioctylphosphine oxide, they are detrimental to efficient electronic transport in quantum dot thin films because of larger interdot spacing and low film density. As such, it has become commonplace to post-synthetically exchange long ligands with short, conductive ones to produce denser, more compact thin films with increased charge transport between individual quantum dots. Numerous studies have shown the impact of varying the length of the final capping ligand on dot packing. Issues with poor surface passivation and limited charge transport in CQDs have lead to strong attention on formation of a secondary material, shell: core/shell heterostructures which can also allow for tunabilty and improved optoelectronic properties. Additionally in many cases the passivating ligands easily desorb from the surface.[22]

4.1.2 Heterostructured type I/II core/shell. Heterostructured core/shell CQDs are composed of one material acting as a shell around another core material and their properties are based on the relative conduction and valence band edge alignment of the core and shell.[24] Typically these are made with a semiconductor-semiconductor core/shell material to improve the photoconductivity efficiency, decrease charge carrier response time, enhance photoluminescence and increase photo-stability. Of particular interest is the use of binary and ternary chalcogenide semiconductor alloys. The inorganic semiconductor shell enables complete and long lasting isolation of the core atoms from the surrounding environment.[22] The choice of the core and shell materials depends on two parameters: band offset alignment and the lattice mismatch between the core and shell.

There are essentially three types of core/shell quantum dots; *Type I, reverse Type I and Type II.*[24] In type I quantum dot heterostructures, the electrons and holes are localized within the core since the shell layer has a higher bandgap and both the conduction and valence band edges of the core are located within the energy gap of the shell. As a result, the emission energy, $h\omega_{PL}$, is determined by the energy gap of the core material, as illustrated in Fig. 5.[22] In essence, the shell is used to more effectively than the capping ligand passivate the surface of the core with the goal of improving its optical properties, separating the more optically active core surface from its surrounding environment and increasing its stability against photo-bleaching. In these however, the shell thickness is critical since the shell layer reduces the material surface activity of the core surface which also decreases photoluminescence quantum yield. For example, the wide-band-gap semiconductor ZnS has become one of the unique shells used for coating of a diversity of group II–VI and III–V quantum dots. The wider-bandgap shell material is being employed as a physical barrier against oxidation, which prevents the core from the surrounding medium effects, and provides better quantum confinement. Reverse type I, is the opposite of type I, where the shell material is of narrower bandgap with electrons confined in the shell. As a consequence, the emission wavelengths can be tuned by changing the thickness of the shell.

In type II heterostructures, one charge carrier is localized in the shell while the other is localized in the core since both the valence and conduction bands of the core are either lower or higher than those in the

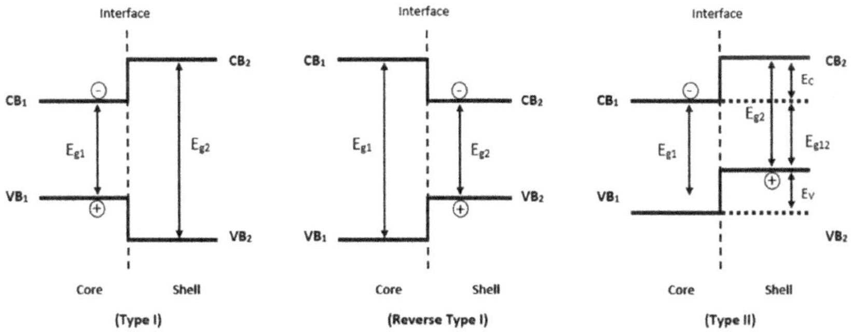

Fig. 5 The three types of core–shell nanocrystals. The upper and lower edges represent the upper and lower energy edges of the core and the shell. (Adapted from ref. 24).

shell. This results in an energy gradient at the interface which spatially separates the electrons and holes on different sides of the heterojunction as shown in Fig. 5. As a consequence, type II quantum dots exhibit longer exciton decay times than type I and can allow access to wavelengths that would otherwise be unavailable for a single material, making them more suitable for photovoltaic applications. In these, the bandgap (E_{g12}) is determined by the energy separation between the conduction band edge of one semiconductor and the valence band edge of the other semiconductor according to eqn (4):[24]

$$E_{g12} = E_{g1} - E_v = E_{g2} - E_c \tag{4}$$

where, E_{g1} and E_{g2} are the bandgaps of semiconductors 1 and 2, respectively, E_c and E_v are conduction and valence band energies, respectively. Unlike type I, tunability of optical properties is achieved through changing thickness of both layers since the bandgap is dependent on both materials and photoluminescence decay times are higher. Since they are CQDs, the passivating ligands prevent agglomeration but the extent of capping does not impact the optical properties.

In order to maximise the benefits of a passivating shell lattice mismatch between core and shell materials must be minimal. When the shell material adapts the lattice parameters of the core during growth, coherency strains result and can play an important role in the optical properties of these core/shell systems. For example Chen *et al.*[25] reported for CdSe/CdS core/shell quantum dots that strain may cause the absorption and emission spectra to be red-shifted. Around the same period, Bawendi and co-workers[26] reported that the average radiative and non-radiative lifetimes of CdTe/CdSe type II core/shell QDs with QY of 4% are 120 times and 3.6 times larger than those of CdTe QDs, respectively. They speculated that the low QY is because the long radiative lifetimes allow for non-radiative channels to dominate. However, when a ZnTe layer is grown onto CdTe/CdSe QDs the QYs increase up to 20%. This they suggest is because the band offsets between CdSe and ZnTe help contain the electron within the dot, without affecting the hole's wavefunction, which is still constrained to the core. Recently, Stride and co-workers[27] reported the preparation of CdSe(S)/Fe_2O_3 core/shell QDs and showed a red-shift in the emission wavelength despite the crystal structure of the core remaining intact after coating with ferric oxide, indicating a direct influence of the shell on the electronic levels of the QD core. The effects of the shell on the optical properties of core/shell QDs of these type-II CdSe(S)/Fe_2O_3 QDs are rationalised by comparison to type-I CdSe(S)/ZnO QDs. Also recently, Wu and co-workers[28] presented comprehensive ultrafast transient absorption and time-resolved PL decay results of quasi-type II CIS/CdS core/shell QDs and demonstrated their improved light harvesting properties over CIS core only QDs. They showed that trapping states in CIS core are effectively passivated by the CdS shell and the single exciton state lifetime was prolonged to 450 ns as a result of reduced electron–hole overlap in quasi-type II CIS/CdS QDs. Likewise, Auger recombination of multiple excitons was suppressed and the bi-exciton lifetime was extended to 42 ps in CIS/CdS QDs from 10 ps in CIS QDs.

Overall, core/shell CQDs are particularly attractive for solar cell applications since the system can be designed to capture energy both in the visible and NIR part of the solar spectrum reducing energy losses and increasing photoconductive efficiency. The core/shell structures discussed are most typical but much research is now being conducted on multi-shell quantum dots such as CdSe/CdS/ZnS,[29,30] CdSe/ZnSe/ZnS,[29,30] InP/ZnSe/ZnS,[31] CdSe/ZnS/CdS/ZnS,[32] which are projected to be even more suited for adapting the solar spectrum in CQDs.

5 Doped CQDs

Unlike size dependent properties, electronic doping in CQDs remains a fairly unexplored area. Despite some successes, many of the early efforts have failed, for reasons that remain unclear.[33] However, there has been some recent advances made in synthesis opening avenues for developments in this area. In heterovalent doping, impurities in a different valence state than the host cations are intentionally incorporated to provide either extra electrons (n-type) or extra holes (p-type). These carriers can introduce extrinsic conductivity to an otherwise poorly conducting material. Recent results suggest that the prototypical semiconductor NCs, CdSe quantum dots, have now been doped n- and p-type through the use of indium, tin, aluminum, or silver. Meanwhile, n- and p-type doping has been definitively shown to occur through noble metal incorporation in InAs NCs.[34] The introduction of dopants can also broaden the range of luminescence properties achievable beyond what is possible in pure materials. Cu and Mn-doped ZnSe NCs can be, for example, alternative, less toxic light emitters compared to CdSe NCs. Tunable dual-colour emission can be achieved in doped heterostructures such as $Zn_{1-x}Mn_xSe$/ZnCdSe core/shell and $Zn_{1-x-y}Cd_xMn_ySe$ alloyed NCs. It is expected that upconverting NCs represent an interesting class of luminescent doped NCs.[34]

Eletronic doping could improve the conductivity of photovoltaic cells constructed of CdSe QDs through excess charge carriers that could fill up trap states. It has been reported that doping has a sigificant impact on the photovoltaic efficiency in PbS QD solar cells. Also nitrogen doping of CdSe QD-sensitized nanocrystalline TiO_2 film solar cells can impact the photoconversion efficiency where the absorption spectrum of the TiO_2 was extended into the visible spectrum indirectly impacting charge separation efficiency.[35] Recently it was reported that core/shell $CuInS_2$–Mn doped CdS QD-sensitized solar cells were assembled and a power conversion efficiency of 5.38% achieved under AM1.5 (100 mW cm^{-2}) illumination.

Though great advancement has been made with doping ions into NCs there remains some critical issues that need still need to be pursued with the use of more powerful characterization techniques. These include, assessing and controlling the location and distribution of dopants in individual NCs and among an ensemble, and developing chemical approaches that can enable a statistically uniform distribution of dopants among the NC population and understanding how the physical properties are influenced by the dopant arrangement.[34] To that end, recent progress towards achieving heavy n-doping of mercury chalcogenides QDs and precisely controlling the number of excess electrons has allowed for the fabrication of the first CQD-based photodetectors in the mid-IR region by exploiting intraband photoexcitations instead of the more conventional interband ones.[36]

6 CQD synthesis

6.1 Nucleation, growth, control of morphology and dispersity

Since the bandgap and consequently the optical properties are size- and shape-dependent, it is an imperative to have a high level of control over these and attain a narrow size distribution (monodispersity) and non-agglomeration. Of the various methods utilized to synthesize quantum dots including *"Top-down"* such as electron-beam lithography and *"Bottom-up"* approaches including Molecular Beam Epitaxy (MBE) and Metal Organic Chemical Vapor Deposition (MOCVD), *colloidal synthesis*, a self-assembly process has emerged as the most reliable, effective, facile and cost-effective in making high quality, phase pure, monodispersed, chemically stable quantum dots that are also tunable in terms of size, shape and optoelectronic properties. This method fundamentally consists of two distinct/separate stages, (1) nucleation of nanoparticles from a solution of high concentration of monomers, and (2) their subsequent growth into particles of specific size and shape.[3] The morphology of the nanoparticle is strictly dependent on ensuring a separation of these stages as when not accomplished result in polydispersity. The method easily facilitates adjustment of reaction parameters including surface passivation *via* ligands or core/shell formation, reaction temperature, time and the nature of ligands and precursors employed. Reactions can be aqueous or organic phases however, organic based approaches tend to produce more monodispersed, non-agglomerated nanoparticles, primarily because the passivation ligands are long hydrocarbon surfactant ligands such as thiols and amines.

The process illustrated in Fig. 6[37,38] typically involves the use of precursor compounds that decompose with temperature resulting in a burst of monomers. Nucleation occurs when a small number of monomers overcome an energetic barrier and assemble into ultra-small clusters/nuclei which are thermodynamically unstable due to high surface energy/chemical potential. The growth of particles involves the successive addition of layers of monomer species onto the surface of the nuclei. There is a critical particle size that is attained relative to the concentration of monomers which dictate the growth rate that is also temperature dependent as summarised:[39]

$$\frac{\mathrm{d}r}{\mathrm{d}t} = \frac{2\sigma\beta C_\infty}{N^2 k_\mathrm{B} T}\left(\frac{1}{r_c} - \frac{1}{r}\right) \tag{5}$$

where, r and r_c are crystal and critical radii, respectively. In this case if $r > r_c$ the crystal grows (the reverse is true) and if they are equal there is no

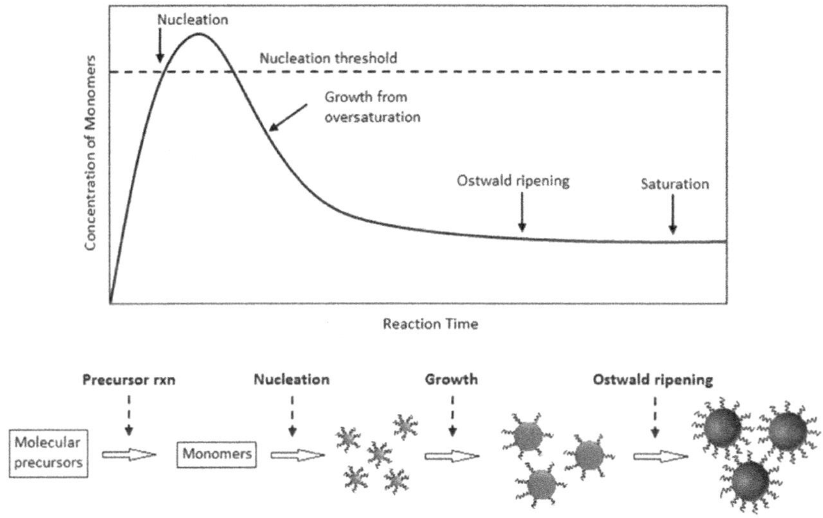

Fig. 6 Illustration of different stages of colloidal synthesis of quantum dots. Top: Reproduced with permission from M. I. Bodnarchuk and M. V. Kovalenko, in *Colloidal Quantum Dot Optoelectronics and Photovoltaics*, ed. G. Konstantatos and E. H. Sargent, Materials Research Society, Cambridge, 2013, pp. 1–29.

growth. Consequentially, smaller crystals shrink and dissolve and larger crystals grow broadening the size distribution in a process called *Ostwald ripening*.[40,41] In effect the target is to grow smaller crystals faster than larger crystals, which is achieved by keeping monomer concentration high by repeated injection of precursors, resulting in narrow size distribution. Additionally, the nucleation event should be quenched and the immediately ensuing growth stage achieved through reduction of temperature and injection of additional precursors.

The use of stabilizing capping/surface active ligands, which bind to the surface of the nanoparticles are critical in helping to control growth and particle shape since the surface is unstable and certain crystal planes preferentially grow. The nature of capping ligands is important since it is used to control the size of nanoparticles. Typically, bulky ligands provide sufficient steric hindrance producing smaller nanoparticles while compact ligands give the reverse effect.[42,43] Also, more bulky ligands tend to produce nanoparticles that are not agglomerated, however as previously indicated they tend to provide an insulating barrier affecting charge transfer mechanisms. Thus the length of the ligands used in nanoparticle synthesis is an important factor when one considers its application in optoelectronic devices where the efficient tunnelling (charge transport) from one particle to the next is critical for optimum performance.[44,45]

The first report of colloidal synthesis was in 1993 by Bawendi and co-workers[46] involving the synthesis of high-quality CdSe nanocrystals by mixing $Cd(CH_3)_2$ and trioctylphosphine selenide, TOPSe in trioctylphosphine oxide, TOPO (a high boiling, weakly polar and coordinating solvent) at 230–260 °C. Since then, this method has been successfully adopted and widely used to synthesize other semiconductor nanocrystals, such as binary CdS, CdTe, ZnSe, ZnS, PbS, PbSe, PbTe, InAs, InP[47,48] as well as multinary I–III-VI and I–II–IV–VI chalcogenides such as $CuInS_2$, $CuInSe_2$ (CIS/Se) and Cu_2ZnSnS_4 and $Cu_2ZnSnSe_4$ (CZTS/Se).[21,43] With early work by O'Brien and co-workers, the use of single precursor in colloidal synthesis has become very attractive.[43,49] These precursors contain both metal and corresponding chalcogen elements and they can decompose at a certain temperature to generate semiconductor compounds directly. With ligands existing in the reaction system, the size and shape of the semiconductor nanocrystals can be controlled. The key point for the success of this method is the selection of a suitable precursor. Clusters and coordination compounds are favourable choices because the central metal ions and ligands can provide all the atoms that the semiconductor compound requires. For example, O'Brien and co-workers[49] obtained CdSe nanocrystals by using the precursor $[CH_3CdSe_2CN(C_2H_5)_2]_2$ decomposing in hot TOPO. Likewise, Strouse and co-workers[50] used $[M_{10}Se_4(SPh)_{16}]_4$ (where M = Cd, Zn) to synthesize CdSe and ZnSe nanocrystals in hexadecylamine solvent.

7 Thin film fabrication

It is important to deposit CQDs onto solid substrates in order to fabricate solar cells or any other electronic devices based on CQDs. Accordingly, a number of physical and chemical methods have been developed to get stable, uniform CQDs films across a large surface area. Among these spin-coating, dip-coating and drop-casting are widely adopted.

7.1 Physical methods

7.1.1 Spin-coating. It is a simple, reproducible method for the deposition of CQDs onto different substrates. In this process, the substrate on which the dots are to be deposited is usually attached to a rotatable vacuum stage. Quantum dots are dispersed in a suitable solvent and deposited on the substrate whilst it is rotating typically at about 1000–5000 rpm. The high speed rotation of substrates with droplets of CQDs allows for even distribution of the QDs. By repeating the deposition and spinning process several times, it is quite easy to increase the thickness of the film.[51–53] High quality films with reproducible thickness are often reported in the literature by this method. Despite being a simple method for producing high quality QD films, spin coating wastes a large portion of QD solution and is sensitive to solvent, speed, humidity and temperature. The main drawback is that scalability is limited.[54]

7.1.2 Drop-casting. Drop-casting is another simple CQD deposition method, in which the quantum dots solution is dropped on a substrate and allowed to dry naturally or with heating.[55] Typically, drop-casting is carried out when QDs are dissolved in high boiling polar solvents like dimethyl formamide or sulfoxide. Such solubility is possible only when the native organic long-chain ligands are effectively exchanged with shorter polar ligands on the surface of QDs.[56] The boiling point of the solvent in this process has profound effect on the rate of evaporation and thereby characteristics of the films.

7.1.3 Dip-coating. Dip-coating method operates using a different mechanism from spin-coating and drop-casting methods. In dip-coating, the substrate is dipped into and withdrawn from a QD solution at a constant rate. Viscosity and surface tension of the solution are major factors that make QDs adhere to the substrate.[57] Though this process wastes negligible amount of QD solution the uniformity of the film is often lacking.

The processes discussed in the foregoing have been explored for research scale fabrication of solar cells and thus far the record efficient CQD solar cell was fabricated following one of the processes. Though they are simple and inexpensive, all the methods severely suffer from scalability. As the CQD field is moving beyond the research scale, large scale production methods are investigated and in this light, ink-jet printing and spray coating are posing to be front runners.[58,59]

7.2 Chemical processing

Unlike with other chemical molecules, effective thin film preparation of quantum dots depends on the characteristics of the surface ligands that passivate dangling bonds. In order to increase the electronic transport in the QD film and to reduce trap density, QDs have to be brought together in contact while maintaining the surface passivation and colloidal stability.

7.2.1 Ligand exchange. Typical CQDs synthesis methods often yield QDs with long-chain ligands on the surface to maintain the colloidal stability. Unfortunately, these long chain ligands are electrically insulating by increasing the space between each dot. For the fabrication of QD solar cells or any other electronic device these ligands have to be exchanged with shorter and more conductive ones. A large number of studies have shown the effect of ligand exchange on the QDs' packing density and conductivity.[60] A variety of shorter ligands with different functional groups have been investigated, which include bifunctional dithiols, mercaptocarboxylic acids, halides and purely inorganic ligands. The results of investigation showed that the densification of QDs depends on type of ligands.[61] Another study comparing different structures of similar length dicarboxylic acid ligands showed the importance of inter particle spacing and strong electronic coupling.[62,63] A similar study using various amines has shown the near complete removal of all capping agents, that led QDs to form ordered arrangement. Such arrangement has exhibited highly efficient electronic transport with high mobility and conductivity.[64] Nevertheless, solar cells fabricated using such arranged QDs did not produce good efficiency. This failure may be related to ineffective reduction of trap state density using the amines. Thus far, the most successful strategies include, purely atomic ligands such as iodine, bromine or hybrid of atomic ligands and short organic ligands such as mercaptopropionic acid (MPA) or ethanedithiol (EDT).[65] An alternative to solid-state ligand exchange is solution state ligand exchange, in which the colloidal stability is maintained by adjusting the stabilizing solvent from non-polar to polar. For such kinds of ligand exchange, metal chalcogenide complexes (MCC) are one of the preferable ligands.[66] However, solar cells using fully solution exchanged QDs have not shown good efficiency. Ligand exchange has thus far been concentrated on metal centers and the chalcogenide part has largely been ignored. Two types of binding modes have been proposed for the thiol ligands in metal sulfide QDs;[67] crystal bound and surface bound. This kind of passivation could be effective in boosting the efficiency of QD solar cells. The highest efficiency so far reported has employed a p-i-n structure, with EDT-capped p-type PbS QDs and iodine capped intrinsic PbS QDs.[68]

7.2.2 Processing conditions. A limited investigation has been carried out in terms of understanding the effect of thin film processing conditions. Factors like temperature, humidity, oxygen and solvent environment can impact QDs film quality and thereby, the solar cell performance. Recently, it has been reported that films annealed to 120 °C have shown a minimal effect, but above 120 °C, loss of ligands, domain growth and loss of quantum confinement was observed.[69] Similarly, the effect of oxygen during film processing has been reported. Depending on the ligand exchange conditions (in air or inert glovebox) the films were found to be showing p- or n-type characteristics.[70]

8 Electrical properties of colloidal quantum dots

As previously indicated important electronic parameters such as carrier mobility, trap density, doping density and diffusion length have profound effect on the performance of quantum dots solar cells.

8.1 Carrier mobility

The parameter that has a crucial role in the performance of CQDs solar cells is the carrier mobility. It is important to measure mobilities of majority and minority carriers accurately in order to understand and improve the performance of solar cells. A few different techniques have been developed for the determination of carrier mobility in quantum dots solar cells. Carrier extraction by linearly increasing the voltage (CELIV) is one of them, which used to measure hole mobility in p-type films. In CELIV, mobility of the carrier is measured based on the time required for the current to reach maximum.[71] The measured hole mobility was 1.5×10^{-3} $cm^2 Vs^{-1}$. Minority mobility was measured by time of flight (TOF) method. To get a reliable value by this method, the film thickness has to be over a micron. The electron mobility was calculated on the basis of the transit time and bias value, the minority mobility calculated by this method was 2.1×10^{-4} $cm^2 Vs^{-1}$. Despite CELIV method being adopted for minority carrier mobility measurements, it was not applicable for all solar cell device structures. Recently,

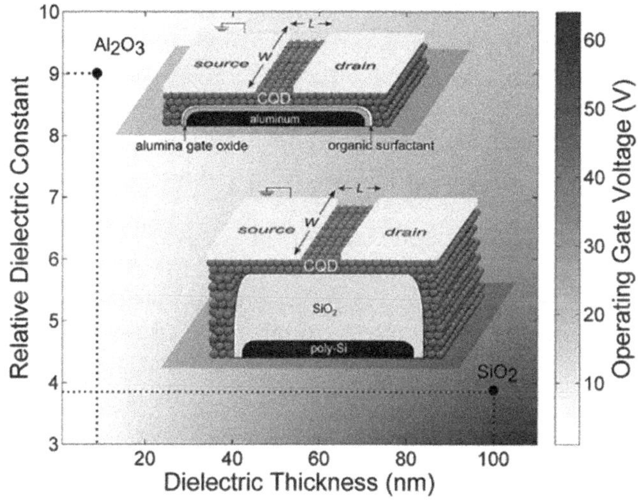

Fig. 7 Schematic representation of a field effect transistor (FET) device structure using a Si/SiO_2 or Al/Al_2O_3 gate. (Reprinted with permission from P. Stadler, B. R. Sutherland, Y. Ren, Z. Ning, A. Simchi, S. M. Thon, S. Hoogland and E. H. Sargent, *ACS Nano*, 2013, **7**, 5757; Copyright 2013 American Chemical Society).

short circuit current density (J_{sc}) transient decay was developed for the determination of minority mobility.[72] Mobility is calculated by recording the transit time (time for the photocurrent to decay to 1/e of its maximum value) over a range of applied biases. Another technique to determine the majority carrier mobility is field effect transistor (FET) as TOF measurement has a stringent thickness requirement. In the FET method, mobility is calculated on the basis of the slope of drain–source current (I_d) *versus* gate voltage. Leakage current has significant effect on the mobility of the carrier determined in this method. In general Si/SiO_2 is used as gate for FET, which has recently been replaced with Al/Al_2O_3 (Fig. 7).[73]

As explained in the previous section surface ligands have profound effect on the mobility of carrier. By replacing the native organic ligand with bromide ions the mobility of the film improved to 10^{-1} $cm^2\,Vs^{-1}$.[74] Interestingly, a similar mobility value has been observed from iodide treated QDs in solid and solution states ($>10^{-1}$ $cm^2\,Vs^{-1}$). From FET measurement, mobility over 10 $cm^2\,Vs^{-1}$ has been observed, but the solar energy conversion efficiency is found to be low.[75]

8.2 Trap density

Uncapped bonds on the surface of quantum dots created during the ligand exchange process induce defects (traps) which can act as a recombination center for the photogenerated carriers and also reduce the quasi-Fermi level splitting range under illumination leading to lower open-circuit voltage. Hence, the reduction of trap density is important for the improvement of solar cell performance. Transient photovoltage decay has been introduced to acquire the trap density in the CQD bandgap.[76] The defect density is calculated on the basis of the integration of the transient current and the ΔV_{oc} value under pulsed light. The type and strength of ligands determine the defect density in quantum dots. Interestingly, iodide ligand is found to show reduced trap density in comparison with organic ligands, the best among the halides.[70] The trap density of 1.8×10^{17} lying 0.36 eV below conduction bands was observed from deep-level transient spectroscopy (DLTS), thermal admittance spectroscopy (TAS) and Fourier transform photocurrent spectroscopy (FTPS) from ethanedithiol capped QDs.[77,78] Other tools such as field effect transistor and photoelectron spectroscopy have also been employed for the determination of trap densities in QDs. Notably, a direct correlation has been observed between photoluminescence QY and solar cell efficiency in luminescent quantum dots, suggesting the trap densities to be the major recombination centers. This observation has led to propose that the highly efficient solar cells should also be high-efficient light emitters.[79]

8.3 Carrier lifetime and diffusion length

Diffusion length of the carrier is another important parameter determining solar cell performance because in actively biased solar cells carrier transport relies mainly on diffusion. Since diffusion length of carrier is determined by the carrier mobility and lifetime, lifetimes also play an important role in performance. To determine the diffusion length of a carrier, a layer of photoluminescence quenching material is added on top of the QD film. By measuring the PL intensity as a function of film thickness, carrier diffusion length can be calculated (Fig. 8).[80] Using the hybrid passivation technique, a diffusion length of up to 70 nm was observed. Recently, with combined dots, fusion and surface passivation a record diffusion length of 230 nm has been achieved.[81]

8.4 Doping density

It has been observed that in PbSe and HgS films that the CQD films processed with lead or mercury cations tend to be n-type, while ones with sulfide or selenide solution are observed to be p-type film.[82] In general n-type films are unstable and they tend to be oxidized in the air. However, developments in solution processing techniques have led to minimize the surface oxidation in n-type films. Band structures of CQD films with different type of ligands were investigated. The investigation showed that the electron accepting ligands make CQD films n-type, whereas electron donating ligands make them p-type.[83] Recently, Zhitomirsky[84] reported doping densities on the order of 10^{16} to 10^{18} cm^{-3} for both p- and n-type films of PbS CQDs. The result of the work demonstrated previously unavailable p-n homojunction within a CQD materials system, and furthermore allows to grade the doping within the active absorber layer to reach power conversion efficiencies (PCEs) exceeding 7%.[84] Lee and co-workers[85] investigated the performance of InAs/GaAs photodetectors as a function of doping density and found that the optimum operation voltage decreases with increasing doping density. It was found that it is necessary to increase the doping density more,

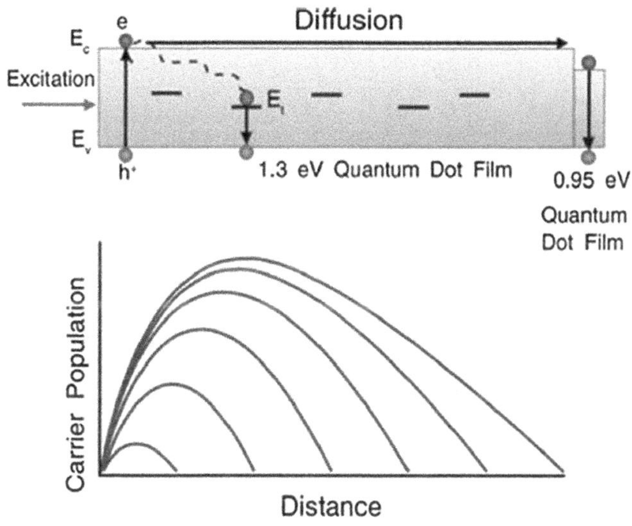

Fig. 8 Schematic representation of diffusion length measurement using the photoluminescence quenching method. The carrier diffusion length was calculated on the basis of the relationship between luminescence intensity and film thickness. (Reprinted with permission from D. Zhitomirsky, O. Voznyy, S. Hoogland and E. H. Sargent, *ACS Nano*, 2013, 7, 5282; Copyright 2013 American Chemical Society).

since the detectivity and responsivity with respect to temperature were improved with increasing the doping density.

9 Solar cell device physics and performance

Recent advances in quantum dots synthesis, surface modification, materials engineering enabled CQDs solar cells to reach energy conversion efficiency close to promising 12%. This section describes various solar cell architectures developed in the past, together with their device physics and performance.

9.1 Schottky CQD solar cells

The concept of using CQDs for the construction of solar cells was tested in the Schottky design.[86] In this, work function difference between indium doped tin oxide (ITO) and magnesium (Mg) top contact was utilized to generate a built in field and QDs and polymer layers were utilized for photocurrent generation. The polymer layers were replaced later with semiconductor nanocrystals layer to form Schottky solar cells, in which CQDs acted as both photo absorber and charge transport medium.[87] Such devices were used as p-type PbS or PbSe CQDs and Al was used as Schottky metal, which was replaced with Ca, Mg or Ag.[88,89] Advancement in ligand strategies, LiF and oxide barrier layers, and organic electron transport layers helped to improve air stability, solar cell performance and efficiency over 5.2%.[90] Fig. 9 shows schematic of a device, absorption spectrum and the current–voltage characteristics of solar cell. Because of easy fabrication and a limited number of interfaces, Schottky devices were attractive as a test bed for studying CQD films

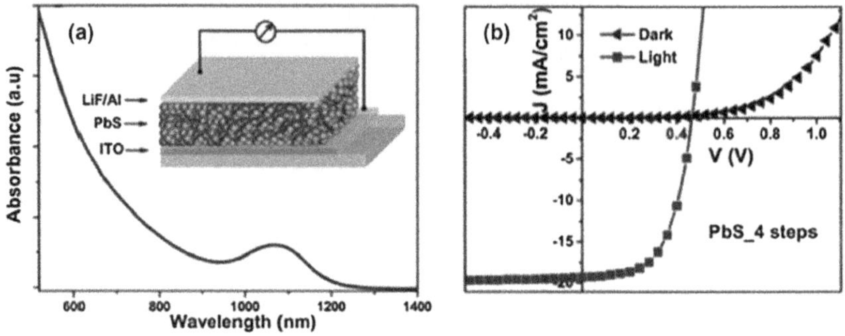

Fig. 9 (a) Absorption spectrum and photovoltaic device structure (inset) for a high-performing PbS nanocrystal Schottky solar cell and (b) Current–voltage characteristics in the dark and under 100 mW cm^2 AM 1.5 illumination of a 5.2% PCE cell with the structure shown in (b).

properties such as band positions of size tuned QDs and work function of metal contacts. However, Fermi level pinning at the interface of metal-CQD reduced the open circuit voltage lower than expected from the CQD bandgap alone.[91] A Major disadvantage of the Schottky design was it required illumination at the nonrectifying side of the junction, which was a problem in transport milted QD films, where it is necessary for maximum photogeneration to occur in the region of maximum internal quantum efficiency close to the junction.

9.2 Heterojunction CQD solar cells

Inspired by the dye-sensitized solar cell design and to address the limitations associated with Schottky devices, heterojunction architecture has been investigated.[92] The design consists of a wide gap n-type mesoporous metal oxide typically TiO_2 or ZnO coated on transparent conducting oxide (ITO or FTO), 50–400 nm thick p-type CQD layer and the stack is finished with a back reflective contact of gold or highly doped oxide followed by silver or aluminum. In this design, the device is illuminated through TCO and the metal oxide, which helps overcomes Schottky limitation. Since the metal oxide layer is heavily doped, majority of the depletion layer falls across the CQDs. Early reports showed excitonic solar cell behavior in this type of architecture; the observation of a depletion layer in the CQDs and a transition to p-n operation was observed upon doping the ZnO side of the junction.[93] The solar energy conversion efficiency of 2.94% was certified from ZnO and PbS nanocrystals device.[94] Ever since this observation, the performance of heterojunction devices have been boosted with developments in engineering design, in particular, by tuning the band positions of metal oxide and quantum dots and ligand strategies towards improving the stability. Interestingly, hybrid passivation strategies such as cadmium and chloride on PbSe showed solar energy conversion efficiency of 6.0%.[65] Nevertheless, heterojunction devices have possessed the absorption–extraction compromise, and it limits the thickness of the absorbing CQD material that can be effectively used in a solar cell.

9.3 Bulk heterojunction CQD solar cells

The limitation of absorption–extraction compromise was addressed using the bulk heterojunction architecture, which is borrowed from organic photovoltaics. This type of architecture ensures that no photogenerated exciton is farther than one diffusion length away from a charge-separating interface. Importantly, minority carrier transport problem is solved by heterojunction design. In bulk heterojunction CQD solar cells, mesoporous metal oxide and CQD film form an interpenetrating layer that improves absorption and carrier collection. An early example of this architecture was a large particle of TiO_2 on top of a thin, compact TiO_2 layer and showed improved light absorption, EQE and photocurrent.[95] Other reports showed lithographically defined TiO_2 nanopillar arrays, hydrothermally grown TiO_2 nanowire network, porous templated TiO_2 wells and ZnO nanowire arrays. Despite addressing some of the problems, bulk heterojunction design has inherent drawback of increase bimolecular recombination, which translates to a lower device built-in voltage.

Another strategy to improve the solar cell efficiency was doping, which was achieved by using different ligands. For example, tetramethylammonium hydroxide was used for the p-type layer and tetrabutylammonium bromide for n-type layer. This doping was implemented in heterojunction architectures. Solar energy conversion efficiency of 7.0% was obtained by lightly doping PbS with 3-mercaptobutyric acid.[96] Similarly, highly doped CuI thin films layer was used as p-type and PbSe QDs as intrinsic layer and ZnO as n-type layer.[97] Air stable solar cells with efficiency of 8.5% were achieved for EDT-treated PbS QDs, in which EDT was used for band bending and for doping (Fig. 10).[98]

9.4 Multi-junction CQD solar cells

The unique characteristics of bandgap tunability with size in quantum dots have been utilized to fabricate multi-junction solar cells. Light

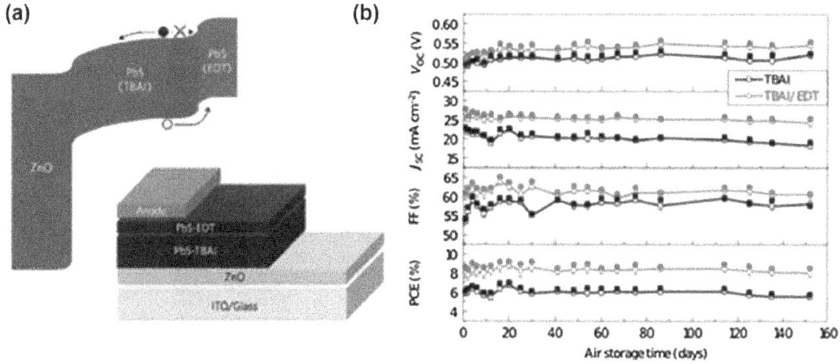

Fig. 10 (a) Schematic representation of band bending at short-circuit conditions in the ZnO/PbS–TBAI/PbS–EDT (TBAI = tetrabutylammonium iodide) device structure. (c) Time evolution of photovoltaic parameters of PbS–TBAI and PbS–TBAI/PbS–EDT devices. Open symbols represent the average values, and solid symbols represent the values for the best-performing device. (Reprinted with permission from Macmillan Publishers Ltd: Nature Materials (ref. 98); Copyright 2014).

absorber material with different bandgaps can be stacked and connected through ideal recombination layers to absorb a wide solar spectrum and to exceed the single junction efficiency limit.[91] However, one of the barriers for this approach is finding a materials system with suitable bandgap tuning. In bulk, thus far, only III–V materials system have been demonstrated for multi-junction solar cells, however the cost associated with materials fabrication limits such devices' potential only to certain high end applications. The cost can be reduced drastically using CQDs materials with two different sizes of a single material that will also address the lattice mismatch problem. The only technical challenge in multi-junction quantum dots solar cells is developing appropriate recombination layers. As a proof of concept, voltage addition has been observed in PbS tandem solar cells using gold islands and graded recombination layer in a depleted heterojunction architecture.[99,100] Further developments in this type solar cells require more research in developing recombination layers.

10 Conclusion and perspective

Constant and collective efforts in colloidal quantum dots research over the past couple of decades have seen tremendous advancements in quantum dots synthesis methods, self-assembly, thin film processing, characterization, solar cells fabrication and theoretical modelling. Such engagements have led to efficiencies of over 10.0% from solar cells entirely based on colloidal quantum dots. However, for the complete realization of quantum dots potential further studies have to be done in every aspect. For instance, to reduce the trap density in quantum dots closer or equal to that observed in bulk semiconductors, better surface passivation strategies have to be developed including new and hybrid ligands. For the enhancement of charge transport in quantum dots films, packing density has to be improved by creating a large area self-assembly process and also impurities on the QDs and in the films have to be reduced if not eliminated. To effectively harness solar energy using QD solar cells, a significant step must be focused towards engineering the band structure of top and bottom contacts, electron and hole transport layers that will have better alignment with the QDs' band structure. Importantly, to raise the solar energy conversion efficiency bar in QD solar cells, multi-junction architecture processing must be matured. Nevertheless, the field of colloidal quantum dot photovoltaics is rapidly growing and certainly has a huge potential to capture a large sector in the solar energy marketplace.

References

1 V. V. Tyagi, N. A. A. Rahim, N. A. Rahim and J. A. L. Selvaraj, *Renewable Sustainable Energy Rev.*, 2013, **20**, 443.
2 M. R. Kim and D. Ma, *J. Phys. Chem. Lett.*, 2015, **6**(1), 85.
3 G. H. Carey, A. L. Abdelhady, Z. Ning, S. M. Thon, O. M. Bakr and E. H. Sargent, *Chem. Rev.*, 2015, **115**, 12732.
4 S. Suresh, *Nanosci. Nanotechnol.*, 2013, **3**(3), 62.

5 K. E. Jasim, *Solar Cells - New Approaches and Reviews*, in L. A. Kosyachenko, InTech, 2015, vol. 11, pp. 303–331.
6 P. Harrison, *Quantum Wells, Wires and Dots*, Wiley, West Sussex, England, 3rd edn, 2009.
7 S. E. White and M. A. Cataluna, *Photonics*, 2015, **2**, 719.
8 L. Brus, *J. Phys. Chem.*, 1986, **90**(12), 2555.
9 PV Education Network online, http://pveducation.org/pvcdrom/solar-cell-operation/short-circuit-current, (accessed September 2016).
10 S. Sagadevan, *Rev. Adv. Mater. Sci.*, 2013, **34**, 44.
11 R. D. Schaller and V. I. Klimov, *Phys. Rev. Lett.*, 2004, **92**, 186601.
12 C. Smith and D. Binks, *Nanomaterials*, 2014, **4**, 19.
13 M. C. Beard, *J. Phys. Chem. Lett.*, 2011, **2**, 1282.
14 O. E. Semonin, J. M. Luther, S. Choi, H.-Y. Chen, J. Gao, A. J. Nozik and M. C. Beard, *Science*, 2011, **334**, 1530.
15 R. J. Ellingson, M. C. Beard, J. C. Johnson, P. Yu, O. I. Micic, A. J. Nozik, A. Shabaev and A. L. Efros, *Nano Lett.*, 2005, **5**(5), 865.
16 S. Tomić, J. M. Miloszewski, E. J. Tyrrell and D. J. Binks, *IEEE J. Photovoltaics*, 2016, **6**(1), 179.
17 A. Teitelboim and D. Oron, *ACS Nano*, 2016, **10**, 446.
18 A. Teitelboim, N. Meir, M. Kazes and D. Oron, *Acct. Chem. Res.*, 2016, **49**, 902.
19 Y. Shang, S. Hao, C. Yang and G. Chen, *Nanomaterials*, 2015, **5**, 1782.
20 Z. Deutsch, L. Neeman and D. Oron, *Nat. Nanotechnol.*, 2013, **8**, 649.
21 M. A. Malik, K. Ramasamy and N. Revaprasadu, in *Nanoscience*, P. O'Brien and P. J. Thomas, Royal Society of Chemistry, Cambridge, 2015, vol. 3, pp. 57–153.
22 J. Du, Z. Du, J.-S. Hu, Z. Pan, Q. Shen, J. Sun, D. Long, H. Dong, L. Sun, X. Zhong and L.-J. Wan, *J. Am. Chem. Soc.*, 2016, **138**, 4201.
23 Y. Altıntas, M. Y. Talpur, M. Ünlü and E. Mutluguün, *J. Phys. Chem. C*, 2016, **120**, 7885.
24 R. G. Chaudhuri and S. Paria, *Chem. Rev.*, 2012, **112**, 2373.
25 X. Chen, Y. Lou, A. C. Samia and C. Burda, *Nano Lett.*, 2003, **3**(6), 799.
26 S. Kim, B. Fisher, H.-J. Eisler and M. Bawendi, *J. Am. Chem. Soc.*, 2003, **125**, 11466.
27 F. Mirnajafizadeh, F. Wang, P. Reece and J. A. Stride, *J. Mater. Sci.*, 2016, **51**(11), 5252.
28 K. Wu, G. Liang, D. Kong, J. Chen, Z. Chen, X. Shan, J. R. McBride and T. Lian, *Chem. Sci.*, 2016, **7**, 1238.
29 D. V. Talapin, I. Mekis, S. Go1tzinger, A. Kornowski, O. Benson and H. Weller, *J. Phys. Chem. B*, 2004, **108**, 18826.
30 B. Omogo, F. Gao, P. Bajwa, M. Kaneko and C. D. Heyes, *ACS Nano*, 2016, **10**, 4072.
31 T. Greco, C. Ippen and A. Wedel, *Proc. SPIE 8424, Nanophotonics IV,* 2012, 842439.
32 P. Linkov, V. Krivenkov, I. Nabiev and P. Samokhvalov, *Mater. Today: Proc.*, 2016, **3**(2), 104.
33 S. C. Erwin, L. Zu, M. I. Haftel, A. L. Efros, T. A. Kennedy and D. J. Norris, *Nature*, 2005, **436**(7), 91.
34 R. Buonsanti and D. J. Milliron, *Chem. Mater.*, 2013, **25**, 1305.
35 V. Chikan, *J. Phys. Chem. Lett.*, 2011, **2**, 2783.
36 A. Stavrinadis and G. Konstantatos, *Chem. Phys. Chem.*, 2016, **17**, 632.
37 C. B. Murray, C. R. Kagan and M. G. Bawendi, *Annu. Rev. Mater. Sci.*, 2000, **30**, 545.

38 M. I. Bodnarchuk and M. V. Kovalenko, in *Colloidal Quantum Dot Optoelectronics and Photovoltaics*, G. Konstantatos and E. H. Sargent, Materials Research Society, Cambridge, 2013, pp. 1–29.
39 M. Booth, PhD thesis, The University of Leeds, 2014.
40 Y. De Smet, L. Deriemaeker and R. Finsy, *Langmuir*, 1997, **13**, 6884.
41 Y. De Smet, L. Deriemaeker, E. Parloo and R. Finsy, *Langmuir*, 1999, **15**, 2327.
42 C. B. Murray, S. Sun, W. Gaschler, H. Doyle, T. A. Betley and C. R. Kagan, *IBM J. Res. Dev.*, 2001, **45**(1), 47.
43 J. Chang and E. R. Waclawik, *RSC Adv.*, 2014, **4**(45), 23505.
44 F. Xu, L. F. Gerlein, X. Ma, C. R. Haughn, M. F. Doty and S. G. Cloutier, *Materials*, 2015, **8**, 1858.
45 S. Bhaumik and A. J. Pal, *J. Phys. Chem. C*, 2013, **117**(48), 25390.
46 C. B. Murray, D. J. Noms and M. G. Bawendi, *J. Am. Chem. Soc.*, 1993, **115**, 8706.
47 Z. Zhuang, Q. Peng and Y. Li, *Chem. Soc. Rev.*, 2011, **40**, 5492.
48 F.-J. Fan, L. Wu and S.-H. Yu, *Energy Environ. Sci.*, 2014, **7**, 190.
49 T. Trindade and P. O'Brien, *Adv. Mater.*, 1996, **8**, 161.
50 S. L. Cumberland, K. M. Hanif, A. Javier, G. A. Khitrov, G. F. Strouse, S. M. Woessner and C. S. Yun, *Chem. Mater.*, 2002, **14**, 1576.
51 M. Shimomura and T. Sawadaishi, *Curr. Opin. Colloid Interface Sci.*, 2001, **6**, 11–16.
52 Y. Fu, H. Xu, S. Bai, D. Qiu, J. Sun, Z. Wang and X. Zhang, *Macromol. Rapid Commun.*, 2002, **23**, 256.
53 T. Cassagneau, T. E. Mallouk and J. H. Fendler, *J. Am. Chem. Soc.*, 1998, **120**, 7848.
54 A. R. Kirmani, G. H. Carey, M. Abdelsamie, B. Yan, D. Cha, L. R. Raollny, X. Cui, E. H. Sargent and A. Amassian, *Adv. Mater*, 2014, **26**, 4717.
55 A. Fischer, L. Rollny, J. Pan, G. H. Carey, S. M. Thon, S. Hoogland, O. Voznyy, D. Zhitomirsky, J. Y. Kim and O. M. Bakr, *Adv. Mater*, 2013, **25**, 5742.
56 Z. Ning, H. Dong, Q. Zhang, O. Voznyy and E. H. Sargent, *ACS Nano*, 2014, **8**, 10321.
57 C.-C. Y. Yang, J. Josefowicz and L. Alexandru, *Thin Solid Films*, 1980, **74**, 117.
58 Q. Guo, S. J. Kim, M. Kar, W. N. Shafarman, R. W. Birkmire, E. A. Stach, R. Agrawal and H. W. Hillhouse, *Nano Lett.*, 2008, **8**, 2982.
59 I. J. Kramer, J. C. Minor, G. Moreno-Bautista, L. Rollny, P. Kanjanaboos, D. Kopilovic, S. M. Thon, G. H. Carey, K. W. Chou and D. Zhitomirsky, *Adv. Mater.*, 2015, **27**, 116.
60 F. Remacle, *J. Phys. Chem. A*, 2000, **104**, 4739.
61 A. H. Ip, S. M. Thon, S. Hoogland, O. Voznyy, D. Zhitomirsky, R. Debnath, L. Levina, L. R. Rollny, G. H. Carey and A. Fischer, *Nat. Nanotechnol.*, 2012, **7**, 577.
62 M. H. Zarghami, Y. Liu, M. Gibbs, E. Gebremichael, C. Webster and M. Law, *ACS Nano*, 2010, **4**, 2475.
63 A. R. Smith, W. Yoon, W. B. Heuer, S. I. M. Baril, J. E. Boercker, J. G. Tischler and E. E. Foos, *J. Phys. Chem. C*, 2012, **116**, 6031.
64 W. J. Baumgardner, K. Whitham and T. Hanrath, *Nano Lett.*, 2013, **13**, 3225.
65 J. Zhang, J. Gao, C. P. Church, E. M. Miller, J. M. Luther, V. I. Klimov and M. C. Beard, *Nano Lett.*, 2014, **14**, 6010.
66 Z. Ning, H. Dong, Q. Zhang, O. Voznyy and E. H. Sargent, *ACS Nano*, 2014, **8**, 1032.
67 M. J. Turo and J. E. Macdonald, *ACS Nano*, 2014, **8**, 20205.
68 C.-H. M. Chuang, P. R. Brown, V. Bulovic and M. G. Bawendi, *Nat. Mater.*, 2014, **13**, 796.

69 B. Ding, Y. Wang, P.-S. Huang, D. H. Waldeck and J.-K. Lee, *J. Phys. Chem. C*, 2014, **118**, 14749.
70 Z. Ning, O. Voznyy, J. Pan, S. Hoogland, V. Adinolfi, J. Xu, M. Li, A. R. Kirmani, J.-P. Sun and J. Minor, *Nat. Mater.*, 2014, **13**, 822.
71 K. W. Johnston, A. G. Pattantyus-Abraham, J. P. Clifford, S. H. Myrskog, S. Hoogland, H. Shukla, E. J. D. Klem, L. Levina and E. H. Sargent, *Appl. Phys. Lett.*, 2008, **92**, 122111.
72 K. Katsiev, A. H. Ip, A. Fischer, I. Tanabe, X. Zhang, A. R. Kirmani, O. Voznyy, I. R. Rollny, K. W. Chou and S. M. Thon, *Adv. Mater.*, 2014, **26**, 937.
73 P. Stadler, B. R. Sutherland, Y. Ren, Z. Ning, A. Simchi, S. M. Thon, S. Hoogland and E. H. Sargent, *ACS Nano*, 2013, **7**, 5757.
74 J. Tang, K. W. Kemp, S. Hoogland, K. S. Jeong, H. Liu, L. Levina, M. Furukawa and X. Wang, *Nat. Mater.*, 2011, **10**, 765–771.
75 S. J. Oh, N. E. Berry, J.-H. Choi, E. A. Gaulding, T. Palik, S.-H. Hong, C. B. Murray and C. R. Kagan, *ACS Nano*, 2013, **7**, 2413.
76 A. H. Ip, S. M. Thon, S. Hoogland, O. Voznyy, D. Zhitomirsky, R. Debnath, L. Levina, L. R. Rollny, G. H. Carey and A. Fischer, *Nat. Nanotechnol.*, 2012, **7**, 577.
77 D. Bozyigit, M. Jakob, O. Yarema and V. Wood, *ACS Appl. Mater. Interfaces*, 2013, **5**, 2915.
78 D. Bozyigit, S. Volk, O. Yarema and V. Wood, *Nano Lett.*, 2013, **13**, 5284.
79 O. D. Miller, E. Yablonovitch and S. R. Kurtz, *IEEE J. Photovoltaics*, 2012, **2**, 303.
80 D. Zhitomirsky, O. Voznyy, S. Hoogland and E. H. Sargent, *ACS Nano*, 2013, **7**, 5282.
81 G. H. Carey, L. Levina, R. Comin, O. Voznyy and E. H. Sargent, *Adv. Mater.*, 2015, **27**, 3325.
82 K. S. Jeong, Z. Deng, S. Keuleyan, H. Liu and P. Guyot-Sionnest, *J. Phys. Chem. Lett.*, 2014, **5**, 1139.
83 D. J. Milliron, *Nat. Mater.*, 2014, **13**, 772.
84 D. Zhitomirsky, PhD thesis, University of Toronto, 2014.
85 U. H. Lee, Y. H. Kang, J. H. Oum, S.-J. Lee, M. Kim, S. K. Noh, Y. D. Jang, D. Lee, H. S. Kim, C. H. Park and S. Hong, *Jpn. J. Appl. Phys.*, 2004, **43**(8A), 5199.
86 S. A. McDonald, G. Konstanatos, S. Zhang, P. W. Cyr, E. J. D. Klem, L. Levina and E. H. Sargent, *Nat. Mater.*, 2005, **4**, 138.
87 A. Maria, P. W. Cyr, E. J. D. Klem, L. Levina and E. H. Sargent, *Appl. Phys. Lett.*, 2005, **87**, 213112.
88 J. P. Clifford, K. W. Johnston, L. Levina and E. H. Sargent, *Appl. Phys. Lett.*, 2007, **91**, 253117.
89 J. M. Luther, M. Law, M. C. Beard, Q. Song, M. O. Reese, R. J. Ellingson and A. J. Nozik, *Nano Lett.*, 2008, **8**, 3488.
90 C. Piliego, L. Protesescu, S. Z. Bisri, M. V. Kovalenko and M. A. Loi, *Energy Environ. Sci.*, 2013, **6**, 3054.
91 C. H. Henry, *J. Appl. Phys.*, 1980, **51**, 4494.
92 A. G. Pattantyus-Abraham, I. J. Kramer and A. R. Barkhouse *et. al.*, *ACS Nano*, 2010, **4**, 3374–3380.
93 S. M. Willis, C. Cheng, H. E. Assender and A. A. R. Watt, *Nano Lett.*, 2012, **12**, 1522.
94 J. M. Luther, J. Gao, M. T. Lloyd, O. E. Semonin, M. C. Beard and A. J. Nozik, *Adv. Mater.*, 2010, **22**, 3704.
95 D. A. R. Barkhouse, R. Debnath, I. J. Kramer and D. Zhitomirsky *et. al.*, *Adv. Mater*, 2011, **23**, 3134.

96 M. Yuan, D. Zhitomirsky and V. Adinolfi *et. al.*, *Adv. Mater.*, 2013, **25**, 1719.
97 D.-K. Ko, P. R. Brown, M. G. Bawendi and V. Bulovic, *Adv. Mater.*, 2014, **26**, 4845.
98 C.-H. M. Chung, P. R. Brown, V. Bulovic and M. G. Bawendi, *Nat. Mater.*, 2014, **13**, 796.
99 X. Wang, G. L. Koleilat, J. Tang, H. Liu, I. J. Kramer, R. Debnath and L. Brzozowski *et. al.*, *Nat. Photonics*, 2011, **5**, 480.
100 J. J. Choi, W. N. Wenger, R. S. Hoffman and Y.-F. Lim *et. al.*, *Adv. Mater.*, 2011, **23**, 3144.

The role of ligands in mediating charge transport in nanocrystalline medium

Sean N. Baxter and P. John Thomas*
DOI: 10.1039/9781782620358-00169

There has been increasing realisation of the important role played by ligands, that adorn the surface of nanocrystals, in determining charge transport in a layer of nanocrystals. The past few years have seen attempts to design ligands that facilitate charge transport by forming a robust surface layer that preserves the order in the thin film, while promoting charge transport. This chapter will briefly examine charge transport mechanisms in nanocrystalline layers and look at families of ligands that have been introduced in the recent past.

1 Introduction

Nanocrystals bridge the gap between molecules that possess distinct individual properties and bulk systems that acquire characteristics from a lattice that extends to infinity. Strikingly, the properties of nanocrystals: chemical, physical, electronic and even thermodynamic are determined not just by the chemical identity but also by the dimension of the crystallites.[1–5] The joint role of size and composition in regulating material properties lies at the heart of nanoscience and technology. Nanocrystallites of metal and semiconductors are vital ingredients in potential devices such as, electrical and electrochemical sensors, solar cell devices,[6,7] plasmonic devices,[8] data storage elements,[3] field effect transistors[9] and light-emitting diodes.[10]

Additionally, ligands that bind to the surface of the crystallites and prevent aggregation are important in determining the physics and chemistry of the particulate systems. For example, the dispersibility of particles in a particular solvent are nearly entirely determined by the ligand shell. Condensed nanocrystalline solids are made up to an appreciable quantitiy of surface species as the crystallites possess high specific surface area. Hence, ligands play a key role in enabling smooth charge transport in nanocrystalline charge transport.

Designing nanocrystalline materials for the future calls for an understanding of the effects of size, ligand shells and charge transport (CT) mechanism on properties. This chapter will initially focus on the CT theories and provide a brief overview of the mechanisms pertinent to this area. Later, more recent work on ligand shells of NC materials and their effects on the CT is highlighted.

2 Coulomb blockade and single electron tunnelling

Charge transport studies on a single conducting nanocrystal surrounded by an insulating ligand shell often exhibit distinct single-electron transport

School of Chemistry, Deniol Road, Bangor University, Bangor LL57 2UW, UK.
E-mail: john.thomas@bangor.ac.uk

effects.[14] These effects are particularly prominent when the nanocrystal is small and the energy required for an electron to tunnel onto the particle is far greater than the thermal energy available (k_bT). This can be described by the orthodox theory.[13,15] The electron can tunnel onto the particle when the electrostatic conditions are favourable; the total energy after the addition of an electron onto the particle is required to be smaller than the energy of the initial state. This change in total energy of an electron for tunnelling can be modelled by the circuit in Fig. 1 and by the following equations.[12,13] The electron transfer from particle to particle(or electrode) is governed by the coulomb charging energy (E_c).

$$\Delta E_1^{\pm}(n) = \frac{e_2}{2C_\Sigma} \pm \frac{e}{C_\Sigma}(ne + Q_0 \mp C_2 V) \tag{1}$$

To tunnel through the first junction.

$$\Delta E_2^{\pm}(n) = \frac{e_2}{2C_\Sigma} \pm \frac{e}{C_\Sigma}(ne + Q_0 \pm C_1 V) \tag{2}$$

To tunnel through the second junction. Where n is the number of electrons on the particle, V is applied voltage, $C_\Sigma = C_1 + C_2$ is the total capacitance of the tunnelling and Q_0 is the fractional residual charge surrounding the particle. At values larger than zero, the tunnelling is suppressed due to the higher final energy than the initial states. This is

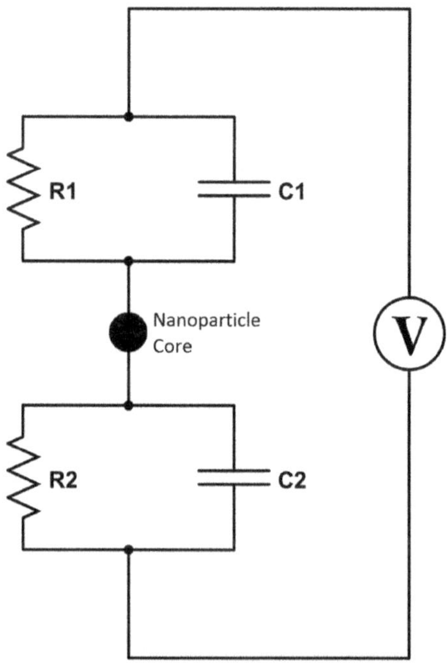

Fig. 1 An equivalent circuit for the double barrier tunnelling junction.This shows the the junction each side of the nanoparticle island. Each junction can be modelled as a resistor and capacitor in parallel.[3,11–13]

referred to as the coulomb blockade.[16] Electrons can only tunnel one at a time when the value for the junction becomes negative.

Observation of the coulomb blockade is temperature dependent as the charging energy needs to be higher than k_bT.

$$E_c > k_bT < \frac{\hbar}{RC_\Sigma} \tag{3}$$

Where R is the resistance of the tunnelling junction. Coulomb blockade can be discerned by non-linear current–voltage curves produced by transport measurements. A typical set of curves exhibiting single electron coulomb blockade as well as staircase behaviour is shown in Fig. 2. The width of the current exclusion zone around nil voltage, called the blockade, depends on the charging energy, dimensionality of the array and the number of elements in the system.[17] The smaller periodic staircase on either side of zero voltage is referred to as the coulomb staircase. It represents the sequential charging of the particle with a single charge carrier as the voltage is scanned across.[18] The coulomb blockade and staircase have both been observed in a number of experiments.[11,13,18–24]

For a single island between source and drain electrodes the resistance and capacitance are related to the distances of the electrodes and ligand shell nature. This leads to a ratio R_1/R_2. where R_2 is believed to be

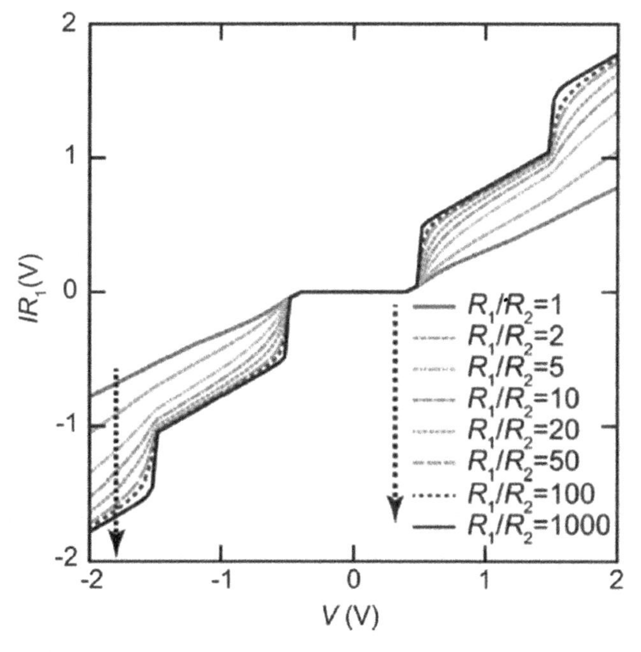

Fig. 2 Set-point current dependence of the whole shape of Coulomb staircases. These curves are simulated by the orthodox theory using $R_1/R_2 = 1, 2, 5, 10, 20, 50, 100$ and 1000, respectively (indicated by the dashed arrow in order). The circuit parameters are $R_1 = 0.1100$ GΩ, $R_2 = 0.1$ GΩ, $C_1 = 0.16$ aF, $C_2 = 0.20$ aF, $Q_0 = 0$e and $T = 4.2$ K. The vertical axis is normalized by multiplying R_1 to compare the whole shape of the Coulomb staircase. A higher R_1/R_2 value gives a sharper Coulomb staircase; in contrast, the shapes of the curves are not dependent on R_1/R_2 values under $R_1/R_2 > 50$. Reproduced from ref. 12 with permission from the Royal Society of Chemistry.

constant due to the ligand molecules and R_1 can be changed by the application of a bias voltage. The shape of any resulting coulomb staircase is strongly dependent on this ratio as shown in Fig. 2.[13,23,24] E_c can be determined directly by scanning tunnelling spectroscopy(STS) measurements.[1,12,23,24]

For ordered arrays, or films, of NCs there are multiple islands for charges to hop through making the phenomenon more complex to model.[19] The value for E_c can be estimated when the NC is thought of as a sphere in a uniform dielectric medium.[25] This provides an estimate in the form of:

$$E_c = \frac{e^2}{4\pi\varepsilon_0\varepsilon_r D} = \frac{e^2}{2C} \quad (4)$$

where, C is the capacitance of the particle, D is the particle diameter and ε_r is the effective dielectric constant for the NC film, this takes into account both the NC and its surrounding medium.[18,22,26] Eqn (4) shows that the E_c is dependent on the size, nature and surrounding ligands for NCs.[27] As an example, a particle with a typical D of 10 nm employing an organic insulating ligand shell, $\varepsilon_r \approx 3.5$ has a charging energy, E_c, of ≈ 40 meV.[18] As the thermal energy at room temperature is approximately, ~25 meV, coulomb charging energy will inhibit the conduction of charge, unless a suitable bias is applied to provide the energy required to overcome the barrier. There exists a voltage threshold V_{th} bellow which the charge transport is uppressed as coulomb blockade dominates. E_c and V_{th} can be related by the following equation

$$V_{\text{th}} = \alpha N E_c \quad (5)$$

where $0 < \alpha < 0.5$ is a value determined by the NC packing and N is the number of islands along the conduction pathway.[28] The dependency of V_{th} on the number of islands in the array means that there will be a decreasing V_{th} with an increase in array size.[20] The visualisation of a coulomb staircase in NC arrays is progressively more difficult as the physical dimensions of the arrays increases.[17] Values calculated for V_{th}, in eqn (5), can be somewhat unreliable as it has also been shown that V_{th} can change with the scale of the I–V measurement.[28]

3 Charge transport in arrays of nanocrystals

There are several theories of the charge transport mechanisms within nanocrystal systems. Generally the medium consists of close packed metal or semiconductor grains held within an insulating matrix.[3] We note that as the dimensions of a solid is reduced, the degeneracy of energy states that characterize the bulk band structure is lifted and the levels are separated by Kubo gap. Such discreetness is particularly pronounced at band edges and could lead to nanocrystal possessing electronic character distinct from the bulk.[29] The electrical properties of granular materials, or in this case nanocrystals, have been extensively studied. The framework was built by Mott,[30] Efros and Shklovskii[31] and

Neugebauer and Webb[32] as well as others. This section shall attempt to provide an overview of the different mechanisms and how they have been recognised in the course of experiments.

3.1 Nearest neighbour hopping

In nearest-neighbour hopping (NNH), conduction occurs mainly through the tunnelling of charge carriers between neighbouring sites. NCs isolated from each other, show a decrease in conductance with temperature,[33,34] fitting the model of activated hopping proposed by Neugebauer and Webb.[32]

$$\sigma \propto e^{-2\delta\beta} e^{-E_a/k_bT} \tag{6}$$

Where δ is the separation between the NCs, β is the tunnelling constant between NCs in a dielectric constant, E_a is the activation energy for charge hopping. In a NC film with a well defined ligand, δ and β may be assumed to be constant, when thermal expansion/contraction is ignored.[14,35] Thus,

$$\sigma = Ae^{-E_a/k_bT} \tag{7}$$

Information of the size of NCs and the effects of the surrounding ligands are contained in the pre-exponential A term. Additionally, size and ligand effects can also influence activation energy.[34] Linear plots of lnσ *vs.* $1/T$ can be used to obtain the activation energy from experimental observations. The activation profile can undergo changes with temperature (see Fig. 3).[33,34]

When E_a is comparable to k_bT, conductivity is nearly independent of temperature, the CT mechanism changes to match the description of the variable range hopping model. Fig. 4 shows such a collapse of E_a in the case of nanocrystalline Si.[36]

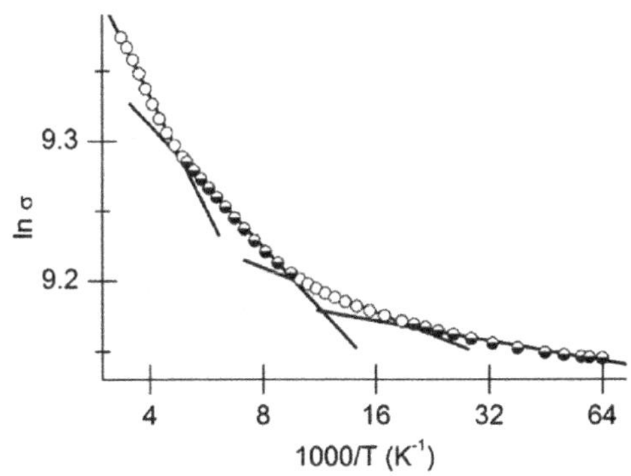

Fig. 3 Plot of lnσ *vs.* $1/T$. The straight lines show four different linear regimes for Activation energy with temperature change. Reprinted with permission from J. Am. Chem. Soc., 2012, 134 (29), 11888–11891. Copyright (2012) American Chemical Society.

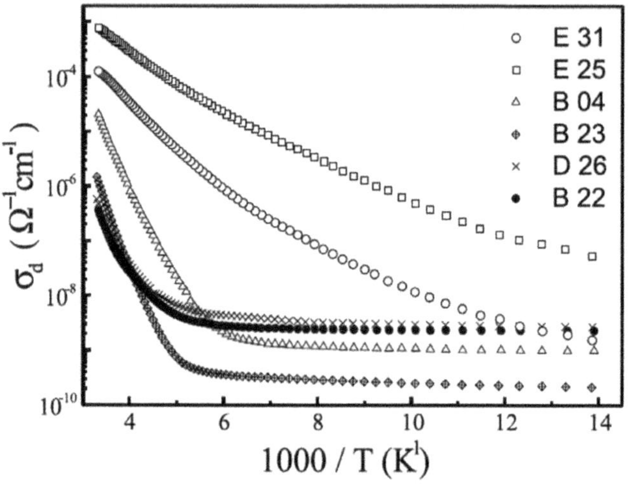

Fig. 4 Conductivity in nanocrystaline silicon films of a range of thickness's (E31 = 1200 nm - B22 = 170 nm). Conductivity appears to be becoming independent of temperature at low temperatures. Reprinted from Journal of Non-Crystalline Solids, 299–302, Ram, Sanjay K. et al, Investigations of the electron transport behavior in microcrystalline Si films, 411–415, Copyright (2002), with permission from Elsevier.

3.2 Variable range hopping

Nanocrystal assemblies within the non-metallic regime, can deviate from the Arrhenius behaviour.[36] The conductivity of these systems can be modelled as:

$$\sigma \approx \exp\left[-\frac{T_0^{\ \nu}}{T}\right] \tag{8}$$

where T_0 is a temperature constant and ν is a characteristic exponent which distinguishes between different mechanisms. If ν is ≈ 1, then NNH Arrhenius behaviour is evident. Variable-range hopping models are used to explains the fractional value of ν. These models assume that charges can acquire the thermal energy required to hop to a new site. Hopping to the next neighbour site is not always energetically favoured since there is a larger probability of finding a more energetically favourable site at a greater hopping distance. The balancing of these two factors, energy difference and distance, are what lead to the fractional value.[35,37] Transport with $\nu = 1/2$[31] and $\nu = 1/3$[38] signify Efros-Shklovskii variable range hopping (ES-VRH) and Mott variable range hopping (Mott-VRH) respectively.[22]

Mott-VRH assumes a constant density of states around the Fermi level, where the hopping conduction occurs between localised states. This allows for the characteristic temperature constant, T_0, for Mott-VRH to be obtained.[39]

$$T_0 \equiv T_M = \left(\frac{3}{k_b N(E_f) a^2}\right) \tag{9}$$

Where $N(E_f)$ is the density of states near the Fermi level and a is the localisation length. However, at low temperatures the density of states is not constant, it vanishes linearly with energy.[39] This results in a coulomb gap that must be overcome by hopping charges. The ES-VRH takes this coulomb gap into account allowing for the energy increase associated with the hopping to overcome the coulomb gap leading to.[39]

$$T_0 \equiv T_E = \left(\frac{2.8e^2}{4\pi\varepsilon\varepsilon_0 k_b a} \right) \tag{10}$$

Where e is the charge of an electron, ε is the dielectric constant of the material and ε_0 is the permittivity of vacuum. Both forms of T_0 are dependent on the localisation length with the ES-VRH also showing a dependence on the dielectric constant of the material. When there is high disorder in the sample the coulomb gap dominates conductance and ES-VRH has been shown to be the preferred hopping method. At samples with low disorder Mott-VRH seams to dominate.[31] It has also been seen that a transition between Mott and ES-VRH can occur when there is a intermediate disorder in the sample.[40]

It is possible to distinguish between the two VRH mechanisms through curve fitting analysis, plotting ln(σ) *vs.* $T^{1/3}$ for Mott-VRH or $T^{1/2}$ for ES-VRH. It has however been reported that data can seemingly fit both regimes.[39] To overcome this ambiguity, the determinant ν can be determined from the following

$$W = -\frac{\delta \ln\rho(T)}{\delta \ln T} = p \times \left(\frac{T_0}{T} \right)^p \tag{11}$$

Where W is the reduced activation energy, hopping energy. Values for p can now be defined by the slope of $\ln W$ *vs.* $\ln T$ since

$$\ln W = \mathrm{A} - p \ln T \tag{12}$$

4 Ligands and charge transport

As shown above, the charge transport properties can be modified through ligand exchange. Altering the ligands surrounding NCs impacts on interparticle distance (D), radius (r) and the effective dielectric constant (ε_r). Ligand exchange reactions allow for surfactants optimised for synthesis to be replaced by those more suited to the final application.[41] The choice of surface ligands control the barriers to electrical transport in NC arrays.[42,43] Many investigations have been undertaken using organic and inorganic ligands to tune the CT properties. Talapin compared the Ligand exchange process to the substitution reaction in coordination chemistry.[41] The types of ligands used have been categorised into 3 groups; donor, L, anion, X, and acceptor, Z (Fig. 5). L type ligands are neutral donor species such as amines and phosphines. X type ligands attach to the NC surface in an anionic manner to terminate the lattice, species such as thiols, carboxylic acids metal free ions[44] and halogens. Finally Z type ligands are neutral acceptor species such as $CdCl_2$ or $AlCl_3$.[41,45–47] Total exchange of surface ligands would be dependent on concentrations of the incoming and outgoing ligands, as well as the

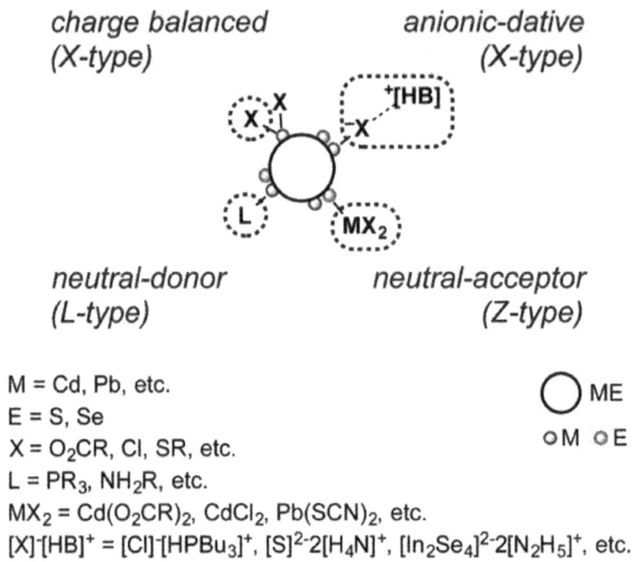

Fig. 5 Ligand binding according to the covalent bond classification. ME represents a NC of binary *II–VI*, *IV–VI*, and *III–V* semiconductors (such as CdS). Reprinted with permission from J. Am. Chem. Soc., 2013, 135 (49), 18536–18548. Copyright (2013) American Chemical Society.

difference in their affinities for the NC surface. For example, dodecanethiol will displace phosphine oxide ligands from Au NCs, however the reverse is unlikely due to the gold–sulfur affinity. This concept can be explained through the hard–soft acid–base principle, where strong electrostatic bonds are formed by hard Lewis acid–base pairs and strong covalent bonds are formed by soft Lewis acid–base pairs.[48,49] Mixtures of hard Lewis acids and soft Lewis base pairs have a poor affinity and do not bind strongly.[41,45] By using ligands which have multiple binding sites such as dithiols and carboxylates the affinity to the NC surface will increase. The steric effects caused by bulky *tert*-butyl ligands when compared to saturated straight chain alkanes will also effect the binding of ligands to the surface, it has been shown that bulky ligands have lower surface coverage when compared to easily packed ligands.[41]

There are two general methods for ligand exchange, in the first method ligands are exchanged in the liquid state. This can be performed in either a one phase system, or a two phase system where the incoming ligand promotes stability in a new medium.[44,47] Liquid phase exchange facilitates creation of devices where the inter NC distances is related to the new ligand.

The second method involves the ligand exchange in solid-state. By soaking NC arrays in solutions containing the incoming ligand, the exchange can be allowed to happen.[33,47,50,51] This allows for a broad range of exchanges, however can also lead to cracking of the layer due to volume contraction caused by replacing large ligands with much smaller ones. This has however been addressed by Luther *et al.*, who used repeated deposition and ligand exchange to attain a smooth layer of a desired thickness.[52]

4.1 Organic ligands

Organic ligands have been investigated in NC arrays have been extensively studied.[8,13,19,33,53,54] Typically, long chain hydrocarbons (C_4–C_{12}) do not allow an efficient charge transfer and transfer of capping ligands with shorter molecules has been practised[55] Investigations have shown that shorter ligands lead to increased conductivities in arrays of alkanethiol-capped Au NCs. Stansfield *et al.* used substituted arylthiol-capped Au NC arrays to demonstrate the substituent effects on the resulting conductivity. The arrays produced in this study showed the ability to be tuned between 100 and 75 000 $S\,m^{-1}$ by simply varying the substituent in the para position.

The *N*-2,4,6-trimethlyphenyl-*N*-methyldithiocarbamate molecule was incorporated onto PbS NCs. The molecule was selected as it is small and conjugated, requiring no further modification in the solid state. The carbodithiolate head is thought to bind strongly to the surface, through a pair of strong metal–sulfur bonds. These NC layers were used in solar cell devices providing, at the time, some of the highest efficiency devices. The ligand also lowers the air sensitivity of the material.[55]

Strikingly, Nakanishi and co-workers were able to produce negative photocurrents in Au and Ag NCs trough ligand modification.[56] Au and Ag NCs of 5.6 ± 0.8 nm diameter were prepared and stabilised with a selection of alkane thiols. The resulting suspension was drop-cast onto glass and evaporated under vacuum. Film thickness of 120–300 nm were reported. The drop-cast films showed a red shift in the surface plasmon resonance band when compared to the suspension, this shift was attributed to aggregation of the NCs.[56] Contacts were sputter coated onto the films and then exposed to white LED light of intensity $I_{white} = 60\ \mu W\,cm^{-1}$. Films comprised of neutral thiols produced a positive current density on irradiation of light. When the films contained charged (either positive or negative) thiols a negative current density was obtained.[57] This negative current density was believed to be due to the trapping of free charge carriers. It is suggested that traps were caused by two things; the NC cores and the organic ligands. The uncharged organic ligands contain no traps and charges tunnel through the insulating ligands. In charged ligands traps exist. These add to those on the NC core. In the dark these traps lay above the Fermi energy and therefore only act as part of the tunnelling barrier. Under irradiation conditions the charge carriers are promoted to higher energies where the traps now have an effect, the density of traps in the system increases. This lowers the number of free carriers leading to a current decrease.[56]

Nanoclusters comprised of 25 gold atoms can be produced with atomic precision using wet chemical techniques.[58] The properties of these nanoclusters can be easily influenced through the use of protecting ligands, in the same way as nanocrystals (Fig. 6). It has been reported that electron-withdrawing ligands favour reduction over oxidation. $[Au_{25}(SNap)_{18}]^-[TOA]^+$ (from herein shown as Au_{25}(SNap)) clusters have been prepared and their electronic properties investigated through UV-vis spectroscopy and DFT calculations.[58] For comparison clusters using Au_{25}(SNap), Au_{25}(SPh) and $Au_{25}(SCH_2CH_2Ph)$ were prepared as solutions

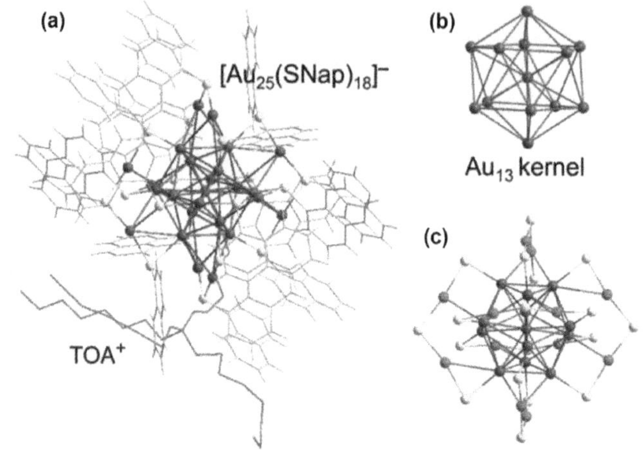

Fig. 6 (a) Total structure of the $[Au_{25}(SNap)_{18}]^{-}[TOA]^{+}$ nanocluster ($Au_{25}S_{18}$ is shown in ball-and-stick mode, thiolate ligands and TOA+ in wire-and-stick mode). (b) Icosahedral Au_{13} kernel. (c) Six dimeric $Au_2(SR)_3$ staple motifs. Gold, green or orange; sulfur, yellow; carbon, gray; hydrogen, white; TOA, magenta. Reprinted with permission from ACS Nano, 2016, 10 (8), 7998–8005. Copyright (2016) American Chemical Society.

in dichloromethane and optical absorption spectra collected. Clusters capped with $Au_{25}(SNap)$, $Au_{25}(SPh)$ were red-shifted to increasing values compared to the $Au_{25}(SCH_2CH_2Ph)$. To understand the red-shifts density functional theory calculations were performed using $Au_{25}(SCH_3)$ and $Au_{25}(SPh)$ as models. The exchange of (SCH_3) with (SPh) caused a slight decrease in the HOMO–LUMO gap, $\approx$0.03 eV, inducing the observed red-shift.[58]

4.2 Inorganic ligands

More recent research has focused on a family of smaller inorganic ligands that promote charge transport and aid particle stability in polar solvents. Unlike organic ligands, these ionic species are widely believed to coordinate strongly to the surface of the NC and provide electrostatic stabilisation in polar solvents.[42,43,59–61] Pronounced changes to charge transport have been reported following the functionalization of the nanocrystals surface with inorganics.

Here, we review detailed studies featuring the following complexes: chalcogenidometallates (also known as metal chalcogenide complexes, MCC) (SnS_4^{2-}, $Sn_2S_6^{4-}$, AsS_3^{3-}, $In_2Se_4^{2-}$),[9,44,61–63] halometallates ($PbCl_3^{-}$, PbI_3^{-}, $FeCl_3^{-}$, $InCl_4^{-}$),[64,65] chalcogenide (S^{2-})[59,66] halides and pseudohalide (Br^{-}, I^{-}, SCN^{-}, N_3^{-})[7,65,67–72] ions.

Pioneering work building on Mitzi's initial reports, which brought MCCs to light,[73] has shown that the MCCs dissolved in anhydrous hydrazine could displace ligands such as: alkyl phosphonic acids(TOPO *etc.*), oleic acid and alkyl thiols[62] commonly used to obtain monodisperse NCs (Fig. 7). Following the ligand exchange, the NCs could move from the non-polar medium into polar solvents such as DMSO, hydrazine.[44,59,61,62] Such a ligand replacement was shown to largely preserve the original optical absorption characteristics of CdSe and Au NCs.[61,62] A striking 10

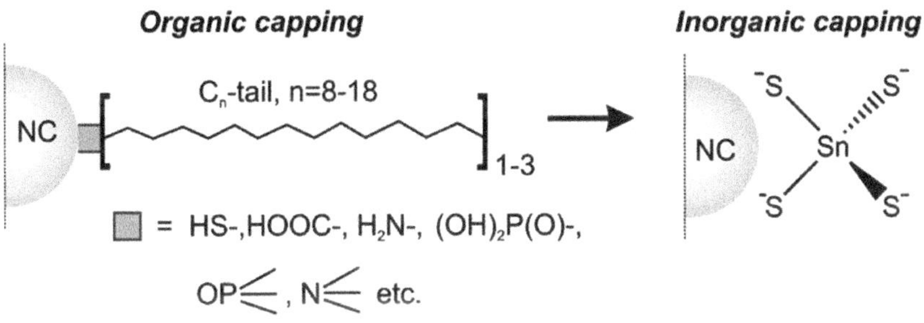

Fig. 7 Scheme showing the ligand exchange process used for preparation of all-inorganic nanocrystals. Reprinted with permission from J. Am. Chem. Soc., 2010, 132 (29), 10085–10092. Copyright (2010) American Chemical Society.

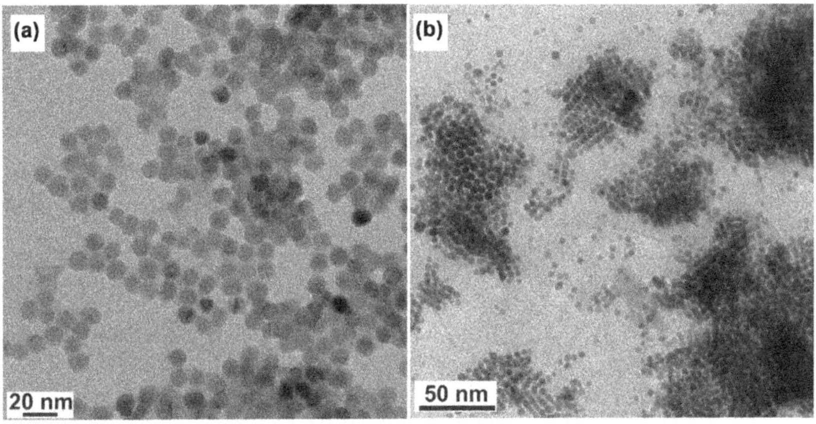

Fig. 8 TEM images of 9.1 nm CdSe, A, and 5 nm Au, B, NCs capped with $Sn_2S_6^{4-}$ ligands. Adapted with permission from J. Am. Chem. Soc., 2010, 132 (29), 10085–10092. Copyright (2010) American Chemical Society.

orders of magnitude improvement in room temperature conductivity following the displacement of the dodecanethiol ligand by $Sn_2S_6^{4-}$ has been observed. The conductivity of, 5 nm, $Sn_2S_6^{4-}$ capped Au increased to *c.a.* 1000 S cm^{-1} from ~10^{-9} S cm^{-1}.[61,62] Reliable SnS_2 shell were obtained on CdSe nanocrystals through the use of the thermally decomposable $(NH_4)_4Sn_2S_6$ molecule (Fig. 8). Whilst in solution the original organic-capped and $(NH_4)_4Sn_2S_6$-capped NCs showed identical absorption spectra. When developed in to solid-state devices the close packed films of MCC-capped CdSe were red shifted, this shift was increased by thermal decomposition at 200 °C:

$$(NH_4)_4Sn_2S_6 \rightarrow 2SnS_2 + 4NH_3 + 2H_2S \quad (13)$$

to leave a very thin shell of SnS_2.[9,62] Field-effect transistors (FET) manufactured with 4.5 nm CdSe/SnS_2 NCs yielded electron mobilities of 1.4×10^{-5} cm^2 V^{-1} s^{-1}.[62] Similar studies on $In_2Se_4^{2-}$-capped CdSe (3.9 nm) and CdSe/CdS (3.6 nm) resulted in values of 10–16 cm^2 V^{-1} s^{-1} and 3.7 cm^2 V^{-1} s^{-1}.

Jang *et al.* performed temperature dependent Hall and Field-effect mobility studies on annealed arrays of 5.3 nm InAs NCs, capped with either $In_2Se_4^{2-}$ or $Cu_7S_4^-$ ligands.[9] They discovered that $In_2Se_4^{2-}$ capped InAs NCs showed a room temperature (RT) Hall mobility of ~1.3 $cm^2 V^{-1} s^{-1}$. However the $Cu_7S_4^-$ capped InAs NCs resulted in a much higher Hall mobility of 16.8 $cm^2 V^{-1} s^{-1}$. This was compared to FET devices manufactured with the same NC suspensions. The $In_2Se_4^{2-}$ capped InAs NCs showed a similar Field-effect mobility, however the $Cu_7S_4^{2-}$ capped InAs NCs showed a lower Field-effect mobility (7.5 $cm^2 V^{-1} s^{-1}$). The difference in mobilities between $In_2Se_4^{2-}$ and $Cu_7S_4^-$ capped InAs was believed to be due to the more efficient tunnelling of electrons through the ligands. The comparison between the Hall and Field-effect results showed that in the low doping system ($Cu_7S_4^-$), the applied gate voltage played a more evident role in the charge transport, than in higher doped systems, by providing a charge accumulation.[9] Building on these results, the Talapin group have proposed the use of MCCs to form composition-matched "solders" for the joining of NCs.[60] They wished to provide alignment between the Fermi levels at the NC ligand boundary to allow efficient CT through the material. This ideal "solder" should be produced of a precursor that, when annealed, would form a semiconducting material which matched the bonded parts. In one example $[Cd_2Se_3]^{2-}$ ions were used to cap 4.7 nm CdSe NCs. Complete replacement of the original *n*-octadecylphosphonic acid ligands was confirmed by infrared (IR) spectroscopy. The $[Cd_2Se_3]^{2-}$-capped CdSe NC suspension was spin-coated onto doped Si and used in FET mobility measurements. Post annealing at 250 °C the electron mobility was measured at 210 $cm^2 V^{-1} s^{-1}$. This at the time exceeded any other reported values for solution-processed inorganic semiconductors.[60] Further annealing at 300 °C produced devices measuring FET mobilities of >300 $cm^2 V^{-1} s^{-1}$, a mobility which compares well to the best bulk solution-processed materials.[43] This was believed to confirm the need for designing "solders" to join NCs which match the energy levels within the materials.

MCC complexes do however have their associated problems. The preparation on the compounds requires inert atmospheres due to their unstable nature.[9,61] It was also identified that the ionic composition differed, dependent on the solvent used. For reasons such as these researchers began looking into simpler, smaller and more stable molecules for ligands.[49,59]

The thiocyanate (SCN^-) ion has been demonstrated as an air stable, environmentally benign, short ligand.[70,71,74,75] SCN^- is suited as it can bind to the metal sites on the surface of the NC and electrostatically stabilise a variety of metal and semiconductor NC materials in polar solvents, such as: PbS, PbSe, CdS, CdSe and CdTe. The ligand was able to reduce the interparticle distance and increase the coupling between particles in films prepared post ligand exchange.[71] Devices produced with SCN^--capped (5–8 nm) Au NCs were recorded to have conductivities of $2 \pm 0.7 \times 10^3$ S cm^{-1} this is an improvement by a factor of almost 10^{12} on the as prepared Au NCs.[71] NC spheres (6.3 nm) and cubes (11 nm) of

PbS were produced and ligand exchanged with the (SCN^-) ion. FET devices employing each of the PbS morphologies and annealed at 120 °C were generated. The cubes showed the higher current of the two, recording electron mobilities of $0.1 \pm 0.04\ cm^2\ V^{-1}\ s^{-1}$, compared to the spheres electron mobility of $0.02 \pm 0.01\ cm^2\ V^{-1}\ s^{-1}$. The cube morphology was also annealed at higher temperatures, 150 °C, producing a device with a much higher electron mobility of $0.33 \pm 0.18\ cm^2\ V^{-1}\ s^{-1}$.[70] Reports using SCN^--capped CdSe NCs have provided a range of electron mobilities (1.5–21.9 $cm^2\ V^{-1}\ s^{-1}$) dependent on the conditions and materials used as the source/drain contacts in FET devices.[71,74,75]

Arrays of SCN^--capped PdSe have been investigated with the addition of $PbCl_2$, in an atomic layer deposition process. This is believed to fill the vacant Pb sites producing a more stable NC.[69,76] Conductivities of 20 $S\ cm^{-1}$ and electron mobilities of up to 7 $cm^2\ V^{-1}\ s^{-1}$ were measured in arrays of 5.9 nm NCs using Au contacts.[69] The treatment with $PbCl_2$ was shown to reduce the hole density whilst increasing the electron density in the NC arrays.[69] The technique was investigated at different temperatures and time scales. A 6 hour treatment at 65 °C produced arrays with electron mobilities of $1.4 \times 10^{-1}\ cm^2\ V^{-1}\ s^{-1}$, and 12 hours yielded an electron mobility of 2.4 $cm^2\ V^{-1}\ s^{-1}$. An alternative method using 95 °C for just 10 minutes produced an array with an electron mobility of $\approx 4.7\ cm^2\ V^{-1}\ s^{-1}$.[76]

Non-chalcogenide, metal halide and halide ions have been employed as inorganic ligands for NC colloids.[7,64,65,67] Halides are simple, short, readily available and offer a reduced toxicity when compared to other inorganic ligands being used. It has been shown that treatment with halide or halide salts leads to a reduction in surface trap sites and a reduced inter NC distance in array devices.[67] Br^--capped PbS NCs in a RT process have achieved electron mobilities of $4 \times 10^{-2}\ cm^2\ V^{-1}\ s^{-1}$, a result comparable to some MCC type ligands after annealing at 200 °C.[7] Reports using I^--capped CdSe NCs show an electron mobility of 12.8 $cm^2\ V^{-1}\ s^{-1}$ after annealing at 200 °C, which could be improved to $\approx 30.8\ cm^2\ V^{-1}\ s^{-1}$ at an annealing temperature of 300 °C.[65] The azide ligand, N_3^-, has also been demonstrated as a suitable ligand for capping InAs NC colloids.[65] FET devices produced using N_3^--capped InAs annealed at 250 °C had electron mobilities of 0.16 $cm^2\ V^{-1}\ s^{-1}$, higher temperature annealing did show a slight improvement, but not to the same scale as the I^--capped CdSe.[65]

Ligand control is offering one of the more versatile routes in producing novel, tuned materials. Careful manipulation can yield changes of many orders of magnitude for electron mobilities and conductivities. Resulting in materials which can, one day, be built into specific devices and function as required. The newer, smaller inorganic ligands are producing arrays with much closer spacing and are thought to improve the mobility of electrons through the materials, as evidenced above. The functionalisation of NC is promising for applications in photovoltaics, FETs, molecular electronics, light-emitting diodes and many other wide ranging applications. We still require full investigations of the limits of sol and solid-state ligand exchanges. Do exchanges with smaller molecules in the solid state produce cracking and breaking up of arrays? Is a solid state exchange as efficient a method for production methods of the future?

References

1 D. V. Talapin, J.-s. Lee, M. V. Kovalenko and E. V. Shevchenko, *Chem. Rev.*, 2010, **110**, 389–458.
2 C. N. R. Rao, P. J. Thomas and G. U. Kulkarni, *Nanocrystals: Synthesis, Properties and Applications*, Springer, Berlin, Heidelberg, 2007, vol. 95, pp. 1–186.
3 A. Zabet-Khosousi and A.-A. Dhirani, *Chem. Rev.*, 2008, **108**, 4072–4124.
4 M. Nasilowski, B. Mahler, E. Lhuillier, S. Ithurria and B. Dubertret, *Chem. Rev.*, 2016, **116**, 10934–10982.
5 K. L. Kelly, E. Coronado, L. L. Zhao and G. C. Schatz, *J. Phys. Chem. B*, 2003, **107**, 668–677.
6 D. B. Mitzi, M. Yuan, W. Liu, A. J. Kellock, S. J. Chey, V. Deline and A. G. Schrott, *Adv. Mater.*, 2008, **20**, 3657–3662.
7 J. Tang, K. W. Kemp, S. Hoogland, K. S. Jeong, H. Liu, L. Levina, M. Furukawa, X. Wang, R. Debnath, D. Cha, K. W. Chou, A. Fischer, A. Amassian, J. B. Asbury and E. H. Sargent, *Nat. Mater.*, 2011, **10**, 765–771.
8 J. Wei, N. Schaeffer and M. P. Pileni, *J. Phys. Chem. B*, 2014, **118**, 14070–14075.
9 J. Jang, W. Liu, J. S. Son and D. V. Talapin, *Nano Lett.*, 2014, **14**, 653–662.
10 Z. Chen, B. Nadal, B. Mahler, H. Aubin and B. Dubertret, *Adv. Funct. Mater.*, 2014, **24**, 295–302.
11 P. J. Thomas, G. U. Kulkarni and C. N. R. Rao, *Chem. Phys. Lett.*, 2000, **321**, 163–168.
12 S. Kano, T. Tada and Y. Majima, *Chem. Soc. Rev.*, 2015, **44**, 970–987.
13 H. Zhang, Y. Yasutake, Y. Shichibu, T. Teranishi and Y. Majima, *Phys. Rev. B*, 2005, **72**, 205441.
14 S. Karmakar, S. Kumar, P. Marzo, E. Primiceri, R. Di Corato, R. Rinaldi, P. G. Cozzi, A. P. Bramanti and G. Maruccio, *Nanoscale*, 2012, **4**, 2311.
15 P. Zhang and T. K. Sham, *Appl. Phys. Lett.*, 2002, **81**, 736.
16 S. Pradhan, J. Sun, F. Deng and S. Chen, *Adv. Mater.*, 2006, **18**, 3279–3283.
17 C. P. Collier, T. Vossmeyer and J. R. Heath, *Annu. Rev. Phys. Chem.*, 1998, **49**, 371–404.
18 J. Kane, J. Ong and R. F. Saraf, *J. Mater. Chem.*, 2011, **21**, 16846.
19 C. Duan, Y. Wang, J. Sun, C. Guan, S. Grunder, M. Mayor, L. Peng and J. Liao, *Nanoscale*, 2013, **5**, 10258–10266.
20 P. Yang, I. Arfaoui, T. Cren, N. Goubet and M. P. Pileni, *J. Phys.: Condens. Matter*, 2013, **25**, 335302.
21 R. Parthasarathy, X.-M. Lin and H. M. Jaeger, *Phys. Rev. Lett.*, 2001, **87**, 4.
22 D. Yu, C. Wang, B. L. Wehrenberg and P. Guyot-Sionnest, *Phys. Rev. Lett.*, 2004, **92**, 216802.
23 S. Kano, Y. Azuma, M. Kanehara, T. Teranishi and Y. Majima, *Appl. Phys. Express*, 2010, **3**, 105003.
24 H. Koo, S. Kano, D. Tanaka, M. Sakamoto, T. Teranishi, G. Cho and Y. Majima, *Appl. Phys. Lett.*, 2012, **101**, 083115.
25 B. Skinner, T. Chen and B. I. Shklovskii, *Phys. Rev. B*, 2012, **85**, 1–13.
26 H. Grabert and M. H. Devoret, *Single Charge Tunneling*, Springer US, Boston, MA, 1992, vol. 294, p. 21.
27 T. Chen, B. Skinner, W. Xie, B. I. Shklovskii and U. R. Kortshagen, *J. Phys. Chem. C*, 2014, **118**, 19580–19588.

28 J. Liao, S. Blok, S. J. van der Molen, S. Diefenbach, A. W. Holleitner, C. Schönenberger, A. Vladyka and M. Calame, *Chem. Soc. Rev.*, 2015, **44**, 999–1014.
29 C. N. R. Rao, G. U. Kulkarni, P. J. Thomas and P. P. Edwards, *Chem. - Eur. J.*, 2002, **8**, 28–35.
30 N. Mott, *Philos. Mag.*, 1972, **11**, 1–18.
31 A. L. Efros and B. I. Shklovskii, *J. Phys. C Solid State Phys.*, 1975, **8**, L49–L51.
32 C. A. Neugebauer and M. B. Webb, *J. Appl. Phys.*, 1962, **33**, 74.
33 G. L. Stansfield and P. J. Thomas, *J. Am. Chem. Soc.*, 2012, **134**, 11888–11891.
34 P. J. Thomas, G. L. Stansfield, N. Komba, D. J. H. Cant, K. Ramasamy, E. Albrasi, H. Al-Chaghouri, K. L. Syres, P. O'Brien, W. R. Flavell, E. Mubofu, F. Bondino and E. Magnano, *RSC Adv.*, 2015, **5**, 62291–62299.
35 K. Whitham, J. Yang, B. H. Savitzky, L. F. Kourkoutis, F. Wise and T. Hanrath, *Nat. Mater.*, 2016, **15**, 557–563.
36 S. K. Ram, S. Kumar, R. Vanderhaghen and P. Roca i Cabarrocas, *J. Non. Cryst. Solids*, 2002, **299–302**, 411–415.
37 P. Guyot-Sionnest, *J. Phys. Chem. Lett.*, 2012, **3**, 1169–1175.
38 N. Mott, *J. Non. Cryst. Solids*, 1968, **1**, 1–17.
39 D. Joung and S. I. Khondaker, *Phys. Rev. B*, 2012, **86**, 235423.
40 H. Liu, A. Pourret and P. Guyot-Sionnest, *ACS Nano*, 2010, **4**, 5211–5216.
41 M. A. Boles, D. Ling, T. Hyeon and D. V. Talapin, *Nat. Mater.*, 2016, **15**, 141–153.
42 C. R. Kagan, E. Lifshitz, E. H. Sargent and D. V. Talapin, *Science*, 2016, 353.
43 R. Wang, Y. Shang, P. Kanjanaboos, W. Zhou, Z. Ning and E. H. Sargent, *Energy Environ. Sci.*, 2016, **9**, 1130–1143.
44 L. Protesescu, M. Nachtegaal, O. Voznyy, O. Borovinskaya, A. J. Rossini, L. Emsley, C. Copéret, D. Günther, E. H. Sargent and M. V. Kovalenko, *J. Am. Chem. Soc.*, 2015, **137**, 1862–1874.
45 N. C. Anderson, M. P. Hendricks, J. J. Choi and J. S. Owen, *J. Am. Chem. Soc.*, 2013, **135**, 18536–18548.
46 M. V. Kovalenko, L. Manna, A. Cabot, Z. Hens, D. V. Talapin, C. R. Kagan, V. I. Klimov, A. L. Rogach, P. Reiss, D. J. Milliron, P. Guyot-Sionnest, G. Konstantatos, W. J. Parak, T. Hyeon, B. A. Korgel, C. B. Murray and W. Heiss, *ACS Nano*, 2015, **9**, 1012–1057.
47 C. R. Kagan and C. B. Murray, *Nat. Nanotechnol.*, 2015, **10**, 1013–1026.
48 P. Reiss, M. Carrière, C. Lincheneau, L. Vaure and S. Tamang, *Chem. Rev.*, 2016, **116**, 10731–10819.
49 A. Nag, H. Zhang, E. Janke and D. V. Talapin, *Z. Phys. Chem.*, 2015, **229**, 85–107.
50 G. L. Stansfield, P. V. Vanitha, H. M. Johnston, D. Fan, H. AlQahtani, L. Hague, M. Grell and P. J. Thomas, *Philos. Trans. R.S Soc., A*, 2010, **368**, 4313–4330.
51 J.-H. Choi, H. Wang, S. J. Oh, T. Paik, P. Sung, J. Sung, X. Ye, T. Zhao, B. T. Diroll, C. B. Murray and C. R. Kagan, *Science*, 2016, **352**, 205–208.
52 J. M. Luther, M. Law, Q. Song, C. L. Perkins, M. C. Beard and A. J. Nozik, *ACS Nano*, 2008, **2**, 271–280.
53 L. Wang, P.-A. Albouy and M. P. Pileni, *Chem. Mater.*, 2015, **27**, 4431–4440.
54 F. Schulz, G. T. Dahl, S. Besztejan, M. A. Schroer, F. Lehmkühler, G. Grübel, T. Vossmeyer and H. Lange, *Langmuir*, 2016, **32**, 7897–7907.
55 R. Debnath, J. Tang, D. A. Barkhouse, X. Wang, A. G. Pattantyus-Abraham, L. Brzozowski, L. Levina and E. H. Sargent, *J. Am. Chem. Soc.*, 2010, **132**, 5952–5953.
56 H. Nakanishi, K. J. M. Bishop, B. Kowalczyk, A. Nitzan, E. A. Weiss, K. V. Tretiakov, M. M. Apodaca, R. Klajn, J. F. Stoddart and B. A. Grzybowski, *Nature*, 2009, **460**, 371–375.

57 S. C. Warren, D. A. Walker and B. A. Grzybowski, *Langmuir*, 2012, **28**, 9093–9102.
58 G. Li, H. Abroshan, C. Liu, S. Zhuo, Z. Li, Y. Xie, H. J. Kim, N. L. Rosi and R. Jin, *ACS Nano*, 2016, **10**, 7998–8005.
59 A. Nag, M. V. Kovalenko, J.-s. Lee, W. Liu, B. Spokoyny and D. V. Talapin, *J. Am. Chem. Soc.*, 2011, **133**, 10612–10620.
60 D. S. Dolzhnikov, H. Zhang, J. Jang, J. S. Son, M. G. Panthani, T. Shibata, S. Chattopadhyay and D. V. Talapin, *Science*, 2015, **347**, 425–428.
61 M. V. Kovalenko, M. I. Bodnarchuk, J. Zaumseil, J.-S. Lee and D. V. Talapin, *J. Am. Chem. Soc.*, 2010, **132**, 10085–10092.
62 M. V. Kovalenko, M. Scheele and D. V. Talapin, *Science*, 2009, **324**, 1417–1420.
63 R. W. Crisp, J. N. Schrauben, M. C. Beard, J. M. Luther and J. C. Johnson, *Nano Lett.*, 2013, **13**, 4862–4869.
64 D. N. Dirin, S. Dreyfuss, M. I. Bodnarchuk, G. Nedelcu, P. Papagiorgis, G. Itskos and M. V. Kovalenko, *J. Am. Chem. Soc.*, 2014, **136**, 6550–6553.
65 H. Zhang, J. Jang, W. Liu and D. V. Talapin, *ACS Nano*, 2014, **8**, 7359–7369.
66 H. Zhang, B. Hu, L. Sun, R. Hovden, F. W. Wise, D. A. Muller and R. D. Robinson, *Nano Lett.*, 2011, **11**, 5356–5361.
67 H. Zhang, J. Yang, J.-R. Chen, J. R. Engstrom, T. Hanrath and F. W. Wise, *J. Phys. Chem. Lett.*, 2016, 7, 642–646.
68 P. R. Brown, D. Kim, R. R. Lunt, N. Zhao, M. G. Bawendi, J. C. Grossman and V. Bulović, *ACS Nano*, 2014, **8**, 5863–5872.
69 S. J. Oh, Z. Wang, N. E. Berry, J.-H. Choi, T. Zhao, E. A. Gaulding, T. Paik, Y. Lai, C. B. Murray and C. R. Kagan, *Nano Lett.*, 2014, **14**, 6210–6216.
70 W.-K. Koh, S. R. Saudari, A. T. Fafarman, C. R. Kagan and C. B. Murray, *Nano Lett.*, 2011, **11**, 4764–4767.
71 A. T. Fafarman, W.-k. Koh, B. T. Diroll, D. K. Kim, D.-K. Ko, S. J. Oh, X. Ye, V. Doan-Nguyen, M. R. Crump, D. C. Reifsnyder, C. B. Murray and C. R. Kagan, *J. Am. Chem. Soc.*, 2011, **133**, 15753–15761.
72 S. J. Oh, N. E. Berry, J.-H. Choi, E. A. Gaulding, T. Paik, S.-H. Hong, C. B. Murray and C. R. Kagan, *ACS Nano*, 2013, 7, 2413–2421.
73 D. B. Mitzi, *Adv. Mater.*, 2009, **21**, 3141–3158.
74 D. K. Kim, Y. Lai, B. T. Diroll, C. B. Murray and C. R. Kagan, *Nat. Commun.*, 2012, **3**, 1216.
75 J.-H. Choi, A. T. Fafarman, S. J. Oh, D.-K. Ko, D. K. Kim, B. T. Diroll, S. Muramoto, J. G. Gillen, C. B. Murray and C. R. Kagan, *Nano Lett.*, 2012, **12**, 2631–2638.
76 S. J. Oh, N. E. Berry, J.-H. Choi, E. A. Gaulding, H. Lin, T. Paik, B. T. Diroll, S. Muramoto, C. B. Murray and C. R. Kagan, *Nano Lett.*, 2014, **14**, 1559–1566.